Memoirs of the American Mathematical Society

Number 423

Shek-Tung Wong

The meromorphic continuation and functional equations of cuspidal Eisenstein series for maximal cuspidal groups

Published by the
AMERICAN MATHEMATICAL SOCIETY
Providence, Rhode Island, USA

January 1990 · Volume 83 · Number 423 (fifth of 6 numbers)

1980 *Mathematics Subject Classification* (1985 *Revision*).
Primary 11F12; Secondary 11F72.

Library of Congress Cataloging-in-Publication Data

Wong, Shek-Tung, 1956–
The meromorphic continuation and functional equations of cuspidal Eisenstein series for maximal cuspidal groups/Shek-Tung Wong.
p. cm. – (Memoirs of the American Mathematical Society, ISSN 0065-9266; no. 423)
"January 1990, volume 83, number 423 (fifth of 6 numbers)."
Includes bibliographical references.
ISBN 0-8218-2486-4
1. Eisenstein series. 2. Automorphic forms. 3. Decomposition (Mathematics) 4. Spectral theory (Mathematics) I. Title. II. Series.
QA3.A57 no. 423 [redacted]
[QA295]
510 s–dc20 [redacted] 89-28399
[515′.243] CIP

Subscriptions and orders for publications of the American Mathematical Society should be addressed to American Mathematical Society, Box 1571, Annex Station, Providence, RI 02901-1571. *All orders must be accompanied by payment.* Other correspondence should be addressed to Box 6248, Providence, RI 02940-6248.

SUBSCRIPTION INFORMATION. The 1990 subscription begins with Number 419 and consists of six mailings, each containing one or more numbers. Subscription prices for 1990 are $252 list, $202 institutional member. A late charge of 10% of the subscription price will be imposed on orders received from nonmembers after January 1 of the subscription year. Subscribers outside the United States and India must pay a postage surcharge of $25; subscribers in India must pay a postage surcharge of $43. Each number may be ordered separately; *please specify number* when ordering an individual number. For prices and titles of recently released numbers, see the New Publications sections of the NOTICES of the American Mathematical Society.

BACK NUMBER INFORMATION. For back issues see the AMS Catalogue of Publications.

MEMOIRS of the American Mathematical Society (ISSN 0065-9266) is published bimonthly (each volume consisting usually of more than one number) by the American Mathematical Society at 201 Charles Street, Providence, Rhode Island 02904-2213. Second Class postage paid at Providence, Rhode Island 02940-6248. Postmaster: Send address changes to Memoirs of the American Mathematical Society, American Mathematical Society, Box 6248, Providence, RI 02940-6248.

10 9 8 7 6 5 4 3 2 1 94 93 92 91 90

ABSTRACT

We carry out, in the context of an algebraic group and an arithmetic subgroup, an idea of Selberg for continuing Eisenstein series. It makes use of the theory of integral operators. The meromorphic continuation and functional equation of an Eisenstein series constructed with a cusp form on the Levi component of a rank one cuspidal subgroup are established. In the Introduction we give a brief account of the role of Eisenstein series in the decomposition of the regular representation.

Received by the editor October 20, 1987. Received in revised form August 2, 1989.

Key words: Eisenstein series, automorphic forms, spectral decomposition.

Prepared at the University of California, Santa Cruz.

TABLE OF CONTENTS

LIST OF GENERAL NOTATIONS

		Page
G	the reductive real algebraic group, Lie algebra $\mathbf{g}$	41
$\tilde{G}$	a real reductive Lie group	52
G_Q	the points of G over $\mathbf{Q}$	42, 170
Γ	the arithmetic subgroup	41
K	a maximal compact subgroup, Lie algebra $\boldsymbol{k}$	42
P_0	the minimal cuspidal subgroup, $P_0 = N_0A_0M_0$, Lie algebras: $\boldsymbol{p}_0, \boldsymbol{n}_0, \boldsymbol{a}_0, \boldsymbol{m}_0$	43
Ξ	a complete set for representatives for $\Gamma \setminus G_Q/(P_0)_Q$	171
P_r	a maximal standard cuspidal subgroup	43
$P_{r\nu_r}$ $(P_{r\nu})$	a G_Q-conjugate of P_r	171 (174)
P	A fixed maximal cuspidal subgroup (except in Section 2.4) P = NAM, Lie algebras: $\boldsymbol{p}, \boldsymbol{n}, \boldsymbol{a}, \boldsymbol{m}$, complex Lie algebra: $\boldsymbol{a}_C$ dual: $\boldsymbol{a}_C^*$	46
P_j (P_i)	a maximal standard cuspidal subgroup associated with P, $P_j = N_jA_jM_j$, Lie algebras: $\boldsymbol{p}_j, \boldsymbol{n}_j, \boldsymbol{a}_j, \boldsymbol{m}_j$ complex Lie algebra: $(\boldsymbol{a}_j)_C$ dual: $(\boldsymbol{a}_j)_C^*$	100 (135)
$W(\boldsymbol{a}, \boldsymbol{a}_j)$	the Weyl group of $(\boldsymbol{a}, \boldsymbol{a}_j)$	100
$W(\boldsymbol{a}, \boldsymbol{a}_{j\nu})$	the Weyl group of $(\boldsymbol{a}, \boldsymbol{a}_{j\nu})$	180
Γ_M	$\pi_M(\Gamma \cap NM)$ $(\pi_M : P \to M)$	44

		Page
K_M	$K \cap M$	44
Σ^o, $\{\alpha_1, \ldots, \alpha_m\}$	the set of simple roots of $(\boldsymbol{p}_0, \boldsymbol{a}_0)$	43
$\eta(u)$	$\inf_{1 \le r \le m} \alpha_r(u) \quad (u \in \boldsymbol{a}_0)$	88
$H_0(g)$	$\log a_0 \in \boldsymbol{a}_0$ for $g = n_0 a_0 k_0$ $(n_0 \in N_0, a_0 \in A_0, k_0 \in K)$	110
$H_j(g)$	$\log a_j \in \boldsymbol{a}_j$ for $g = n_j a_j m_j k$ $(n_j \in N_j, a_j \in A_j, m_j \in M_j, k \in K)$	110
S_0	the Siegel domain	87, 170
F	the fundamental domain in $\Xi\, S_0$ (in Chapters 1-5, the fundamental domain in S_0)	88, 172
$\\| \ \\|$	the usual matrix norm on $G \subset GL(n, \mathbf{R})$	42
α	a kernel function	74
$\alpha * f$	a convolution	47
δ_r	a truncation factor	90
δ_{rv}	a truncation factor	176
λ_α	a smoothing factor	93
$[\ ]_F$	Γ-automorphic extension from F to all of G	93
$\mathbf{K}_\alpha$	a compact operator on $L^2(\Gamma \backslash G)$	93
spec $\mathbf{K}_\alpha$	the spectrum of $\mathbf{K}_\alpha$	122
$\\| \ \\|_2$	the L^2 norm on $L^2(\Gamma \backslash G)$	80
R	the resolvent of $\mathbf{K}_\alpha$	122
Φ	the cusp form used to define the Eisenstein series $E(\Lambda, \Phi, g)$	46

		Page
σ	a character of K	42
$E(\Lambda, \Phi, g)$	the Eisenstein series	46
$E(\underline{\Lambda}, \underline{\Phi}, g)$	the complete Eisenstein series	194, 195
$(\boldsymbol{a}_C^*)^+$	the domain of convergence of the Eisenstein series $E(\Lambda, \Phi, g)$	46
$\boldsymbol{D}(\tilde{G})$	the left $\tilde{G}$-invariant differential operators on $\tilde{G}$	45
$\boldsymbol{D}_K(\tilde{G})$	the left $\tilde{G}$-invariant and right K-invariant differential operators on $\tilde{G}$	45
$\mathbf{Z}(\tilde{G})$, $(\mathbf{Z}(G))$	the bi-invariant differential operators on $\tilde{G}$ (G)	45 (79)
$\chi_\Lambda(D)$	the eigenvalue of $D \in \mathbf{Z}(G)$ with the Eisenstein series $E(\Lambda, g)$	47
$\hat{\alpha}^o(\Lambda)$	a Selberg transform	144
$\hat{\alpha}(\Lambda)$	$(\hat{\alpha}^o(\Lambda))^3$	120
f^r (f^j, f^i)	the constant term f with respect to the rth (j^{th}, i^{th}) maximal standard cuspidal subgroup	44 (142, 167)
f^{rv}	the constant term of f with respect to P_{rv}	175
$M(s \mid \Lambda)$	an intertwining operator	100
$M(\underline{\Lambda})$	thc complete intertwining operator	194, 196
$\phi(j, s, n \mid \Lambda)$, $\phi(J \mid \Lambda)$	a constant term coefficient	101
$H_{j,s,n}(\Lambda, g)$, $H_J(\Lambda, g)$	a function derived from the constant term of the Eisenstein series $E(\Lambda, g)$ with respect to P_j	120
$F^{**}_{j,s,n}(\Lambda, g)$, $F_J^{**}(\Lambda, g)$	a Fredholm solution	120

		Page
S	the complete set of poles of $E(\Lambda, g)$	155
W^c (set theory)	the complement of W	
F' (general topology)	the set of limit points of F	
$\overline{W}$ (general topology)	the closure of W	

INTRODUCTION

An Eisenstein series is typically constructed through summing the translates of a function whose domain is under a discontinuous group action, a formal device for producing invariant functions first employed systematically by Eisenstein to construct the basic elliptic functions (Eisenstein [1], Weil [1]). He introduced the following series. Given a lattice $\Gamma = \{m_1\omega_1 + m_2\omega_2 : m_1, m_2 \in \mathbf{Z}\}$ in the complex plane, define

$$(1) \qquad E_n(z) = \sum_{\gamma \in \Gamma} \frac{1}{(z+\gamma)^n}$$

$$= \sum_{(m_1, m_2) \in \mathbf{Z}^2} \frac{1}{(z + m_1\omega_1 + m_2\omega_2)^n}$$

for a positive integer n. The issue of convergence was well recognized by Eisenstein. For $n \geq 3$, the series is absolutely convergent whenever z is not in Γ, and defines a doubly periodic function meromorphic in the whole plane with poles at the points of Γ. For $n = 1$ and $n = 2$, Eisenstein took the series to be the iterated sum

$$\lim_{B\to\infty} \sum_{m_2=-B}^{B} \lim_{A\to\infty} \sum_{m_1=-A}^{A} \frac{1}{(z + m_1\omega_1 + m_2\omega_2)^n}.$$

Among other things he obtained the Laurent expansion of each E_n at the origin:

$$E_n(z) = \frac{1}{z^n} + (-1)^n \sum_{k=1}^{\infty} \binom{k-1}{n-1} e_k z^{k-n}$$

where, for k even,

$$(2) \qquad e_k = \sum_{(m_1, m_2)}{}' \frac{1}{(m_1\omega_1 + m_2\omega_2)^k} = \sum_{(m_1, m_2) \in \mathbf{Z}^2 \setminus \{0\}} \frac{1}{(m_1\omega_1 + m_2\omega_2)^k} \qquad (k \geq 4),$$

$$e_2 = \lim_{B\to\infty} \sum_{m_2=-B}^{B}{}' \lim_{A\to\infty} \sum_{m_1=-A}^{A}{}' \frac{1}{(m_1\omega_1 + m_2\omega_2)^2};$$

while for k odd, $e_k = 0$. One sees from the definitions (1) and (2) that $E_2 - e_2$ is just the $\wp$-function introduced by Weierstrass a decade and a half later after the cited work of Eisenstein. With the basic elliptic functions thus defined, Eisenstein was able to derive many of the formulae and relations in the theory of elliptic functions through essentially formal developments.

Eisenstein's construction, though first conceived as a mean of producing elliptic functions, allows a quite different interpretation when one shifts one's focus from the dependence on z to the dependence on the pair of periods (ω_1, ω_2) (Lehner [1]). The modular group $SL(2, \mathbf{Z})$ then replaces the lattice as the group of interest. We now let Γ stand for $SL(2, \mathbf{Z})$. Two pairs of periods (ω_1, ω_2) and (ω_1', ω_2') generate the same lattice, and hence the same $E_n(z)$, if and only if they are related through multiplication by a matrix in $\pm\Gamma$. The summation in the definition (2) may be replaced by one over the action of Γ on the upper half-plane $\mathbf{H}$:

$$e_k = \omega_2^{-k} G_k(\tau)$$

with

$$G_k(\tau) = \sum_{(m_1, m_2)}{}' \frac{1}{(m_1\tau + m_2)^k} \tag{3}$$

$$= 2\zeta(k) \sum_{\gamma \in \pm\Gamma_\infty \backslash \Gamma} \frac{1}{(c\tau + d)^k} \qquad (k \geq 4)$$

for

$$\gamma = \begin{pmatrix} a & b \\ c & d \end{pmatrix} \in \Gamma$$

and

$$\Gamma_\infty = \left\{ \begin{pmatrix} 1 & n \\ & 1 \end{pmatrix} : n \in \mathbf{Z} \right\},$$

where τ denotes ω_1/ω_2 and is taken to lie in $\mathbf{H}$. The series in the definition (3) is a modular form for $PSL(2, \mathbf{Z}) = \pm I \backslash \Gamma$ of weight k. Here enters a general device for producing modular forms.

However, a variant of this was used systematically first, namely, in the construction of the classical Poincaré series, where one sums over a full Fuchsian group instead of a collection of coset representatives.

In his analysis of spaces of modular forms for principal congruence subgroups Hecke turned to the Eisenstein-Poincaré construction for concrete examples (Hecke [1]). He called his generalizations "Eisenstein series." They appeared in the cited reference as

$$G_k(\tau; a_1, a_2, N) = \sum_{\substack{m_1 \equiv a_1 \ (\mathrm{mod}\ N) \\ m_2 \equiv a_2}}{}' \frac{1}{(m_1\tau + m_2)^k} .$$

Typical of the spaces Hecke considered is the vector space of all modular forms of a given weight k for the principal congruence subgroup Γ of level N. For $k \geq 3$, and $N > 2$, there are as many linearly independent absolutely convergent Eisenstein series as there are rational cusps in Γ Eisenstein series such that the constant term of its Fourier expansion at τ_i is 1, and at the other cusps 0. He then deduced that the space is the direct sum of the subspace spanned by the Eisenstein series and the subspace of cusp forms. Both of these subspaces are finite dimensional. For $k = 1$ and $k = 2$, the formally constructed Eisenstein series do not converge absolutely. To remedy the situation Hecke introduced the modified, nonholomorphic Eisenstein series

$$\Phi_k(\tau, s) = \sum_{m_1 \equiv a_i \ (N)}{}' \frac{1}{(m_1\tau + m_2)^k \, | m_1\tau + m_2 |^s} \qquad (k = 1, 2;\ \mathrm{Re}(s) > 1 - \frac{k}{2}) .$$

Earlier he had demonstrated through Fourier expansions that these non-τ-holomorphic Eisenstein series $\Phi_k(\tau, s)$ could be analytically continued to meromorphic functions on the whole s-plane, with $s = 0$ as a regular point (Hecke [2, pp. 391-394]). He then took $\Phi_k(\tau, 0)$ to be his example of modular form of weight 1 or 2.

The subject of our study, nonholomorphic Eisenstein series that are eigenfunctions of differential operators, were first introduced by Maass as examples of wave forms in connection with his extension of the Hecke correspondence between Dirichlet series and modular forms to include zeta functions of real quadratic number fields (Maass [1]). Wave forms are automorphic eigenfunctions of the Laplace-Beltrami operator that have polynomial growth towards the cusps. Following Hecke, Maass used the Eisenstein-Poincaré summation to construct Eisenstein series for a principal congruence subgroup Γ :

$$E(\tau, s; a_1, a_2, N) = \sum_{m_i \equiv a_i\,(N)}{}' \frac{y^{\frac{s}{2}}}{|m_1\tau + m_2|^s} \qquad (\mathrm{Re}(s) > 2)$$

for $\tau = x+iy \in \mathbf{H}$, which may be written as

$$\sum_{t \bmod N} d(s, t, N)\, E^*(\tau, s; ta_1, ta_2, N)$$

with

$$d(s, t, N) = \sum_{\substack{tn \equiv 1\,(N) \\ n>0}} \frac{1}{n^s}$$

and

$$E^*(\tau, s; a_1, a_2, N) = \sum_{\substack{m_i \equiv a_i\,(N) \\ (m_1, m_2) = 1}} \frac{y^{\frac{s}{2}}}{|m_1\tau + m_2|^s}$$

$$= N^{\frac{s}{2}} \sum_{\gamma \in \Gamma_a \backslash \Gamma} y(\sigma_a^{-1}\gamma z)^{\frac{s}{2}},$$

where Γ_a is the stabilizer in Γ of the cusp $-a_2 / a_1$, and $\sigma_a \in SL(2, \mathbf{R})$ is such that $\sigma_a\infty = a$ and $\sigma_a^{-1}\Gamma_a\sigma_a = \pm\Gamma_\infty$. Note that $y^{s/2}$ is an eigenfunction of the Laplace-Beltrami operator

$$D = -y^2\left(\frac{\partial^2}{\partial x^2} + \frac{\partial^2}{\partial y^2}\right).$$

By means of Fourier expansions Maass showed that all the Eisenstein series can be analytically continued to meromorphic functions on the whole s-plane, and are regular on the line $\mathrm{Re}(s) = 1$.

For $s = 1+2ir$ with $r > 0$, s playing the role of weight, in analogy with Hecke's result on vector spaces of modular forms the vector spaces of wave forms were decomposed in terms of Eisenstein series and cusp forms. But the case of $r = 0$ was less satisfactory in this regard. It is of interest to note that Maass made use of the Petersson inner product to obtain an eigenbasis for the Hecke operators acting on some subspaces of cusp forms. With hindsight one can see that in this seminal paper of Maass most of the elements of the spectral theory of $L^2(\Gamma \setminus \mathbf{H})$ were already present.

The role of Eisenstein series in the spectral theory of the Laplace-Beltrami operator D on $L^2(\Gamma \setminus \mathbf{H})$ was first recognized by Selberg ([1, 2]) and, at about the same time, independently by Roelcke ([1]). Selberg was primarily interested in establishing a trace formula, his generalization of the Poisson summation formula from classical harmonic analysis. He actually dealt in the generality of a weakly symmetric Riemannian space under the action of a discontinuous group of isometries, Γ, and considered functions which might not be fully automorphic, but transformed with a representation of Γ. In the present discussion we limit ourselves to the case of functions on $\Gamma \setminus \mathbf{H}$. Selberg introduced the fundamental eigenfunction principle that allows one to correspond the spectrum of D with the spectrum of an integral operator on $L^2(\Gamma \setminus \mathbf{H})$ as defined by an automorphic kernel. Such an automorphic kernel $K(z, z')$ is constructed through summing a point-pair invariant kernel over the action of Γ, and has polynomial growth towards the cusps. He noted that the modified kernel $K^*(z, z')$ obtained from $K(z, z')$ by subtracting certain integrals of Eisenstein series over the line $\mathrm{Re}(s) = 1/2$, one for each member of a complete set of inequivalent cusps, was square integrable over $\Gamma \setminus \mathbf{H} \times \Gamma \setminus \mathbf{H}$. (Selberg parametrized Eisenstein series differently from Maass, the present $\mathrm{Re}(s) = 1/2$ is $\mathrm{Re}(s) = 1$ in Maass [1].) Hence the integral operator defined by $K^*(z, z')$ has a discrete spectrum, and a well defined trace. More importantly, this modified integral operator has exactly the same spectrum in $L^2(\Gamma \setminus \mathbf{H})$ as the original. This is so because, as a consequence of the self-adjointness of D, Eisenstein series are orthogonal to its eigenfunctions in $L^2(\Gamma \setminus \mathbf{H})$. Eisenstein series thus appear to be complementary to the discrete spectrum of D in $L^2(\Gamma \setminus \mathbf{H})$.

Roelcke on the other hand was more directly interested in the spectral theory. He employed the methods of operator theory and the spectral theory of Hilbert spaces to investigate the wave forms introduced earlier by Maass. A general Fuchsian group Γ of the first kind was considered. He obtained first, in case Γ has cusps, the spectral decomposition of $L^2(\Gamma \setminus \mathbf{H})$ with the continuous part described in terms of Hellinger integrals. When Γ is a principal congruence subgroup, for which the Eisenstein series were known to have meromorphic continuation up to the line $\mathrm{Re}(s) = 1$ (parametrization as in Maass), he showed that integrals of Eisenstein series over $\mathrm{Re}(s) = 1$ could be used instead to describe the continuous spectrum, yielding for the first time the decomposition of $L^2(\Gamma \setminus \mathbf{H})$ in terms of Eisenstein series and cusp forms. However, although Eisenstein series for a general Fuchsian group of the first kind with cusps were mentioned, Roelcke did not use them in a similar fashion, their meromorphic continuation up to $\mathrm{Re}(s) = 1$ was still unknown to him.

Both Selberg ([3]) and Roelcke ([2]) gave new proofs of the meromorphic continuation of Eisenstein series. These proofs are important in the development of the theory of Eisenstein series because they make no reference to the specific arithmetic nature of Γ, but rest largely on certain basic analytic properties of Eisenstein series. Later we shall briefly describe these proofs.

The need to broaden the underlying domain of the function space $L^2(\Gamma \setminus \mathbf{H})$ from $\mathbf{H}$ to the group $G = SL(2, \mathbf{R})$ itself was already quite apparent in Selberg's work. Selberg applied his trace formula to evaluate the trace of the Hecke operators acting on the space of holomorphic cusp forms for the modular group of a given weight. As a function on $\mathbf{H}$, a modular form of a nonzero weight k is not fully automorphic. Instead, it transforms as

(4) $$f(\gamma z) = (cz + d)^k f(z)$$

for

$$\gamma = \begin{pmatrix} a & b \\ c & d \end{pmatrix} \in SL(2, \mathbf{Z}) .$$

Selberg found it possible to regard f as a fully automorphic function on a new space, which was formed by adjoining a third coordinate θ to the upper half-plane. While all this was well within his theory of weakly symmetric spaces, he had in effect brought in the group G as the domain of definition of the functions. The connection is easily stated. Recall the Iwasawa decomposition of G:

$$G = NAK$$

where

$$N = \left\{ \begin{pmatrix} 1 & x \\ & 1 \end{pmatrix} : x \in \mathbf{R} \right\} ,$$

$$A = \left\{ \begin{pmatrix} y^{1/2} & \\ & y^{-1/2} \end{pmatrix} : y > 0 \right\} ,$$

$$K = \left\{ \begin{pmatrix} \cos\theta & \sin\theta \\ -\sin\theta & \cos\theta \end{pmatrix} : \theta \in \mathbf{R} \right\} .$$

For every $g \in G$ we have $g = n(x)a(y)k(\theta)$, where $n(x) \in N$, $a(y) \in A$, $k(\theta) \in K$. With regard to the action of G on $\mathbf{H}$ we may correspond $n(x)a(y)$ in G/K with $z = x+iy$ in $\mathbf{H}$. Thus

$$G/K \approx \mathbf{H} .$$

Every f(z) satisfying the transformation rule (4) gives rise to a function f(g) on G:

$$f(g) = f(n(x)a(y)k(\theta)) = y^{k/2} f(z) e^{ik\theta} ,$$

which is easily seen to be automorphic. Under this correspondence holomorphy translates into an eigenfunction property with respect to the Casimir differential operator on G. Moreover it is clear that f(g) transforms with a character of K upon right multiplication by elements of the maximal compact subgroup. The spectral theory of $L^2(\Gamma \setminus \mathbf{H})$ can thus be subsumed within that of $L^2(\Gamma \setminus G)$. The function f(g) is, what is now called, after Harish-Chandra ([1]), an automorphic form on the semi-simple Lie group $G = SL(2,\mathbf{R})$. Automorphic forms have their genesis in Maass' wave forms.

One should however note that, preceding Selberg's work, and almost simultaneous with Maass' paper, the spectral theory of $L^2(\Gamma \setminus G)$ had already been under study by Fomin and Gelfand ([1]) from a representation theoretic point of view. Though there the spectral theory of $L^2(\Gamma \setminus G)$ was pursued for the sake of investigating geodesic flows on manifolds of constant negative curvature, already an interesting connection with modular forms, one concerning the occurrence of the discrete series in the decomposition of the regular representation of G on $L^2(\Gamma \setminus G)$ into irreducible parts, was recognized. Later, after Selberg's work, Gelfand ([1]) was to propose a representation theoretic framework for the theory of automorphic forms, taking under its wings the important results of Selberg, and perhaps a large part of the theory of modular forms as well. Yet Eisenstein series did not figure explicitly in these writings of Gelfand. The first exposition of the use of Eisenstein series to describe the continuous spectrum of the regular representation was given by Godement ([1]), for the case of the modular group.

The spectral decomposition of the regular representation of G on $L^2(\Gamma \setminus G)$, to the extent of the introduction of Eisenstein series to analyze the continuous spectrum, was treated in the context of a real reductive algebraic group G and an arithmetic subgroup Γ by Langlands ([1]). The analytic skeleton of the general development is discernible already in the $\mathrm{rank}_{\mathbf{Q}}(G) = 1$ case, the case of $G = SL(2, \mathbf{R})$ for instance, a brief account of which we now give. Our presentation is modelled on Arthur [1]. One may also refer to Godement [1], Kubota [1], and Lang [1] for more details.

Let Γ be a Fuchsian group of the first kind, i.e., a discrete subgroup of G such that $\Gamma \backslash G$ has finite volume. Since we are primarily interested in the role of Eisenstein series, we suppose that $\Gamma \backslash G$ is noncompact. The reduction theory of such subgroups is well known, and we shall not need to require that Γ be arithmetic. For convenience $P = \pm NA$, the subgroup of upper triangular matrices in G, is supposed to be cuspidal with respect to Γ. In other words, we assume that $N \cap \Gamma \backslash N$ is compact. We also assume that there is a Siegel domain S associated with P that contains a fundamental domain for $\Gamma \backslash G$. Thus

$$G = \Gamma S$$

for

$$S = \Omega_N A(Y) K ,$$

where

$$A(Y) = \left\{ \begin{bmatrix} y^{1/2} & \\ & y^{-1/2} \end{bmatrix} : y > Y > 0 \right\} ,$$

and Ω_N is a compact subset of N. In classical language this means that Γ, when regarded as acting on the upper half-plane, has only one cusp, and which is at ∞.

We begin the spectral decomposition of the right regular representation of G on $L^2(\Gamma \backslash G)$ by introducing an invariant subspace which is the orthogonal complement of the subspace of cusp forms. The elements of this subspace, which are automorphic functions on G, are constructed by means of the Eisenstein-Poincaré summation and, as we shall see, can naturally be expressed in terms of Eisenstein series. For every smooth compactly supported function ψ on $N(\Gamma \cap P) \backslash G$ we define

$$E_\psi(g) = \sum_{\gamma \in \Gamma \cap P \backslash \Gamma} \psi(\gamma g) \qquad (g \in G) .$$

It can be shown that each E_ψ belongs to $L^2(\Gamma \backslash G)$. We let L be the closure of the span of all such functions. There is then the primary decomposition:

$$L^2(\Gamma \setminus G) = {}^{o}L^2 \oplus \mathcal{L}, \tag{5}$$

where ${}^{o}L^2$ is the subspace of cusp forms, i.e., the set of functions in $L^2(\Gamma \setminus G)$ such that the constant term

$$\begin{aligned} f^{o}(g) &= \int_{N \cap \Gamma \setminus N} f(ng)\, dn \\ &= 0 \qquad\qquad \text{for almost all } g\,. \end{aligned}$$

The proof of the primary decomposition is based on the following formal idea:

$$\begin{aligned} \langle E_\psi , f \rangle &= \int_{\Gamma \setminus G} E_\psi(g)\overline{f(g)}\, dg = \int_{\Gamma \setminus G} \sum_{\Gamma \cap P \setminus \Gamma} \psi(\gamma g)\cdot\overline{f(g)}\, dg \\ &= \sum_{\Gamma \cap P \setminus \Gamma} \int_{\Gamma \setminus G} \psi(\gamma g)\, \overline{f(g)}\, dg = \int_{\Gamma \cap P \setminus G} \psi(g)\overline{f(g)}\, dg \\ &= \frac{1}{2} \int_{N \cap \Gamma \setminus G} \psi(g)\overline{f(g)}\, dg \\ &= \frac{1}{2} \int_{N \cap \Gamma \setminus N} \int_A \int_K \psi(ak)\, \overline{f(nak)}\, e^{-\log y}\, dk da(y) dn \\ &= \frac{1}{2} \int_A \int_K \psi(ak) e^{-\log y} \int_{N \cap \Gamma \setminus N} \overline{f(nak)}\, dn dk da(y) \\ &= \frac{1}{2} \int_A \int_K \psi(ak)\, \overline{f^{o}(ak)} e^{-\log y}\, dk da(y)\,, \end{aligned} \tag{6}$$

where for convenience we have assumed that $\Gamma \cap P = \pm\, N \cap \Gamma$. On account of standard harmonic analysis on the compact subgroup K, there is no loss of generality, insofar as the generation of $\mathcal{L}$ is concerned, to suppose that $\psi(g)$ is of the form $\psi(y)e^{-im\theta}$ ($m \in 2\mathbf{Z}$), with $\psi(y)$ being a smooth compactly supported function on A. Let $\mathcal{L}_m$ stand for the closed subspace spanned by all E_ψ constructed with such functions for a given $m \in 2\mathbf{Z}$. Then

$$\mathcal{L} = \bigoplus_{m \in 2Z} \mathcal{L}_m\,.$$

We now analyze the subspaces L_m in terms of Eisenstein series. First, a $\psi(y)$ is expressed in terms of its Mellin transform $\hat{\psi}(z)$:

$$\psi(y) = \frac{1}{2\pi i} \int_{Re(z)=z_0} \hat{\psi}(z) y^{\frac{1+z}{2}} dz \qquad (y > 0), \tag{7}$$

where

$$\hat{\psi}(z) = \int_0^\infty \psi(y) y^{\frac{-(1+z)}{2}} \frac{dy}{y} \qquad (z \in \mathbf{C}),$$

and $z_0 \in \mathbf{R}$ is arbitrary. Then $E_\psi(g)$ can be given as

$$\sum_{\Gamma \cap P \backslash \Gamma} \frac{1}{2\pi i} \int_{Re(z)=z_0} \hat{\psi}(z) y(\gamma g)^{\frac{(1+z)}{2}} e^{-im\theta(\gamma g)} dz,$$

which equals

$$\frac{1}{2\pi i} \int_{Re(z)=z_0} \sum_{\Gamma \cap P \backslash \Gamma} \hat{\psi}(z) e^{-im\theta(\gamma g)} y(\gamma g)^{\frac{(1+z)}{2}} dz,$$

provided the summation and the integration may be interchanged. Let us therefore introduce the Eisenstein series

$$E(z, \Psi(z), g) = \sum_{\gamma \in \Gamma \cap P \backslash \Gamma} \Psi(z, \gamma g) y(\gamma g)^{\frac{(1+z)}{2}}, \tag{8}$$

with

$$\Psi(z) = \Psi(z, g) = \hat{\psi}(z) e^{-im\theta}.$$

It is clear that

$$E(z, \Psi(z), g) = \hat{\psi}(z) E_m(z, g)$$

where

$$E_m(z, g) = \sum_{\gamma \in \Gamma \cap P \backslash \Gamma} e^{-im\theta(\gamma g)} y(\gamma g)^{\frac{(1+z)}{2}}.$$

One can show that $E_m(z, g)$, and hence $E(z, \Psi(z), g)$, is absolutely convergent for $Re(z) > 1$. The foregoing derivation may then be justified, and we have the following integral formula for E_ψ in terms of Eisenstein series:

$$E_\psi(g) = \frac{1}{2\pi i} \int_{Re(z)>1} E(z, \Psi(z), g)\, dz . \tag{9}$$

A basic integral formula for L_m can be deduced from the formula (9). For E_ψ, $E_{\psi'} \in L_m$, we have

$$\begin{aligned}
\langle E_\psi, E_{\psi'} \rangle &= \int_{\Gamma\backslash G} E_\psi(g)\, \overline{E_{\psi'}(g)}\, dg = \int_{\Gamma\backslash G} \frac{1}{2\pi i} \int_{Re(z)>1} E(z, \Psi(z), g)\, \overline{E_{\psi'}(g)}\, dzdg \\
&= \frac{1}{2\pi i} \int_{Re(z)>1} E(z, \Psi(z), g) \int_{\Gamma\backslash G} \overline{E_{\psi'}(g)}\, dgdz \\
&= \frac{1}{2\pi i} \int_{Re(z)>1} E(z, \Psi(z), g) \int_{\Gamma\backslash G} \sum_{\Gamma\cap P\backslash\Gamma} \overline{\psi'(\gamma g)}\, dgdz \\
&= \frac{1}{2\pi i} \int_{Re(z)>1} \int_{\Gamma\cap P\backslash G} E(z, \Psi(z), g)\, \overline{\psi'(g)}\, dgdz \\
&= \frac{1}{4\pi i} \int_{Re(z)>1} \int_A \int_K E^o(z, \Psi(z), ak)\, \overline{\psi'(y)}\, e^{im\theta} \frac{dk(\theta)da(y)dz}{y} .
\end{aligned}$$

Now a well known computation shows the constant term of the Eisenstein series to be

$$E^o(z, \Psi(z), g) = \Psi(z) y^{\frac{(1+z)}{2}} + M(z)\Psi(z) y^{\frac{(1-z)}{2}} ,$$

where $M(z)$ is a holomorphic function of z defined for $Re(z) > 1$, and does not depend on the ψ in consideration. Then

$$\langle E_\psi, E_{\psi'} \rangle = \frac{1}{4\pi i} \int_{Re(z)>1} \int_K \Psi(z)\overline{\Psi'(-\overline{z})} + M(z)\Psi(z)\, \overline{\Psi'(\overline{z})}\, dkdz . \tag{10}$$

To proceed further with our analysis we need to call upon the meromorphic continuation and functional equation of the Eisenstein series $E(z, \Psi(z), g)$, and that of $M(z)$ as well. Bear in mind that $E(z, \Psi(z), g)$ is $E_m(z, g)$ multiplied with an entire function. It can be established that, for any fixed $g \in G$, $E_m(z, g)$, and hence $E(z, \Psi(z), g)$ itself, can be analytically continued to a meromorphic function on the whole z-plane, and is regular on the line $Re(z) = 0$. Likewise for $M(z)$. The poles of $E_m(z, g)$ and $M(z)$ in the region $Re(z) > 0$ all lie in the interval $(0, 1]$, and are finite in number as g varies over G. Moreover, these poles are all simple. Respectively $E_m(z, g)$ and $M(z)$ satisfy the functional equations:

$$E(z, \Psi(z), g) = E(-z, M(z)\Psi(z), g)\,,$$

$$M(z)M(-z) = 1\,.$$

Later we shall enter into these fundamental results on Eisenstein series. Presently we take them for granted.

Now let $\mathcal{L}_m'$ be the closed subspace of $\mathcal{L}_m$ spanned by all E_ψ such that $E(z, \Psi(z), g)$ is regular for $Re(z) \geq 0$. If $E_\psi \in \mathcal{L}_m'$, $M(z)\Psi(z)$ is also regular for $Re(z) \geq 0$. The path of integration in the formula (10) may then be shifted to the line $Re(z) = 0$, and the inner product formula be put in a symmetric form:

$$\begin{aligned}
\langle E_\psi, E_{\psi'} \rangle &= \frac{1}{4\pi i} \int_{Re(z)=0} \int_K \Psi(z)\overline{\Psi'(-\bar{z})} + M(z)\Psi(z)\overline{\Psi'(\bar{z})}\, dkdz \\
&= \frac{1}{4\pi i} \int_{Re(z)=0} \int_K \Psi(z)\overline{\Psi'(z)} + M(z)\Psi(z)\overline{\Psi'(-z)}\, dkdz \\
&= \frac{1}{8\pi i} \int_{Re(z)=0} \int_K \Psi(z)\overline{\Psi'(z)} + M(z)\Psi(z)\overline{\Psi'(-z)}\, dkdz \\
&\qquad + \frac{1}{8\pi i} \int_{Re(z)=0} \int_K \Psi(-z)\overline{\Psi'(-z)} + M(-z)\Psi(-z)\overline{\Psi'(z)}\, dkdz \\
&= \frac{1}{8\pi i} \int_{Re(z)=0} \int_K [\, \Psi(z) + M(-z)\Psi(-z)\,]\, [\, \overline{\Psi'(z)} + \overline{\overline{M(z)}\Psi'(-z)}\,]\, dkdz\,,
\end{aligned}$$

where for the last equality we have used the functional equation of M(z). Furthermore, as $M(z)\overline{M(z)} = 1$ for Re(z) = 0, we finally have

$$(11)\quad \langle E_\psi , E_{\psi'} \rangle = \frac{1}{8\pi i} \int_{Re(z)=0} \int_K [\, \Psi(z) + M(z)^{-1}\Psi(-z)\,]\, \overline{[\Psi'(z) + M(z)^{-1}\Psi'(-z)\,]}\, dkdz\,.$$

Let H be the Hilbert space of all square integrable functions on K, and $\hat{L}$ be the vector space of all measurable functions $F : i\mathbf{C} \to H$ such that

$$\int_{Re(z)=0} \| F(z) \|_K^2 \, dz < \infty\,.$$

The space $\hat{L}$ becomes a Hilbert space with the inner product

$$\langle F_1, F_2 \rangle = \int_{Re(z)=0} \langle F_1(z), F_2(z) \rangle_K \, dz\,.$$

We have

$$\hat{L} = \bigoplus_n \hat{L}_n$$

where each $\hat{L}_n$ consists of functions of the form $f(z)e^{-in\theta}$ ($n \in \mathbf{Z}$). By virtue of the inner product formula (11) there is a unitary map from the closed subspace $\bigoplus_m L'_m$ into the subspace $\bigoplus_m \hat{L}_m$ of $\hat{L}$ as determined by

$$E_\psi \mapsto \Psi(z) + M_m(z)^{-1}\Psi(-z)\,.$$

(Here we make explicit the dependence of M(z) on m.) The image of $\bigoplus_m L'_m$ under this map is of the form $\bigoplus_m \hat{L}'_m$ with $\hat{L}'_m \subset \hat{L}_m$. Now

$$L^2(\Gamma \backslash G) = {}^0L^2 \bigoplus_m L'_m \bigoplus_m L''_m\,,$$

where L''_m is the orthogonal complement of L'_m in L_m. We shall see that $\bigoplus_m L'_m$ is an invariant subspace of the regular representation and that, through the use of Eisenstein series, the equivalent action of G on $\bigoplus_m \hat{L}'_m$ can be described quite explicitly.

For $E_\psi \in \mathcal{L}'_m$, the path of integration in the integral formula (9) may be shifted to the line $\mathrm{Re}(z) = 0$:

$$(12) \qquad E_\psi(g) = \frac{1}{2\pi i} \int_{\mathrm{Re}(z)=0} E(z, \Psi(z), g)\, dz .$$

More generally, for $E_\psi \in \mathcal{L}_m$, we can write

$$\begin{aligned} E_\psi(g) &= \frac{1}{2\pi i} \int_{\mathrm{Re}(z)=0} E(z, \Psi(z), g)\, dz \\ &\quad + \frac{1}{2\pi i} \left[\int_{\mathrm{Re}(z)>1} E(z, \Psi(z), g)\, dz - \int_{\mathrm{Re}(z)=0} E(z, \Psi(z), g)\, dz \right] \\ &= E'_\psi(g) + E''_\psi(g) , \end{aligned}$$

where

$$\begin{aligned} E'_\psi(g) &= \frac{1}{2\pi i} \int_{\mathrm{Re}(z)=0} E(z, \Psi(z), g)\, dz \\ &= \frac{1}{2\pi i} \int_{\mathrm{Re}(z)=0} \hat{\psi}(z) E_m(z, g)\, dz , \end{aligned}$$

and

$$\begin{aligned} E''_\psi &= \sum_j \mathrm{Res}(\hat{\psi}(z) E_m(z, g); z_j) \\ &= \sum_j \hat{\psi}(z_j) \mathrm{Res}(E_m(z, g); z_j) \end{aligned}$$

with j indexing the finite set of poles of $E_m(z, g)$ in the region $\mathrm{Re}(z) > 0$. Since all $\mathrm{Res}(E_m(z,g); z_j)$ belong to $L^2(\Gamma \backslash G)$, so do E'_ψ and E''_ψ. In fact, as we shall see, $E'_\psi \in \mathcal{L}'_m$ and $E''_\psi \in \mathcal{L}''_m$.

First, to show that E'_ψ, and hence also E''_ψ, belongs to the orthogonal complement $\mathcal{L}$ of ${}^0L^2$, we call upon a basic analytic property of the Eisenstein series $E_m(z, g)$. Namely, for every z at which it is regular, the Eisenstein series is an eigenfunction of the Casimir differential operator

ω on G. It can then be deduced from the self-adjointness of ω that E'_ψ is orthogonal to every cuspidal automorphic form. As the set of such cusp forms is dense in ${}^0L^2$, we have $E'_\psi, E''_\psi \in L$. It then follows that $E'_\psi, E''_\psi \in L_m$. Next we note that every $\mathrm{Res}(E_m(z, g); z_j)$ is an eigenfunction of ω with growth $\sim y^{(1-\sigma_j)/2}$, where $\sigma_j = \mathrm{Re}(z_j) > 0$. Also, for $\mathrm{Re}(z) = 0$, $E_m(z, g)$ grows like $y^{1/2}$. From this we deduce, again appealing to the self-adjointness of ω, that $E''_\psi \in L''_m$.

Actually L''_m is precisely the span of the set of residues of $E_m(z, g)$ in the region $\mathrm{Re}(z) > 0$. For with use of the Mellin inversion formula a function E_{ψ_j} in L_m can be constructed for each pole z_j such that $\hat{\psi}_j(z)$ has values 1 at z_j, 0 at the other poles of $E_m(z, g)$. Now, for arbitrary $E_\psi \in L_m$, we have

$$E_\psi = E_{\psi'} + E_{\psi''}$$

where

$$\psi' = \psi - \sum_j \hat{\psi}(z_j)\psi_j, \qquad \psi'' = \sum_j \hat{\psi}(z_j)\psi_j .$$

Since $\hat{\psi}'(z_j) = 0$ for all j, $E_{\psi'} \in L'_m$. On the other hand, $\hat{\psi}''(z_j) = \psi(z_j)$ for all j, and

$$\langle E_{\psi''}, \mathrm{Res}(E_m(z, g); z_j) \rangle = \hat{\psi}(z_j) \| \mathrm{Res}(E_m(z, g); z_j) \|^2 .$$

If u is an element in L_m that is orthogonal to all $\mathrm{Res}(E_m(z, g); z_j)$, and $\{E_{\psi_i}\}$ is a sequence in L_m converging to u, then it must be that $\hat{\psi}_i(z_j) \to 0$ for all j. Hence $E_{\psi''_i} \to 0$, and $E_{\psi'_i} \to u$. This shows that $u \in L'_m$, and establishes our claim that L''_m, the orthogonal complement of L'_m in L_m, is spanned by the residues of $E_m(z, g)$ in the region $\mathrm{Re}(z) > 0$. Finally, as E'_ψ is then orthogonal to all functions in L''_m, we have $E'_\psi \in L'_m$.

We are now in a position to discuss the action of G on $\bigoplus_m L'_m$. For a given $g_0 \in G$, let $T_{g_0}f$ denote the right g_0-translate of a function f on G. For $E_\psi \in L'_m$,

$$T_{g_0}E_\psi(g) = \frac{1}{2\pi i} \int_{\mathrm{Re}(z)=0} E(z, \Psi(z), gg_0)\, dz$$

$$= \frac{1}{2\pi i} \int_{\mathrm{Re}(z)=0} \hat{\psi}(z) E_m(z, gg_0)\, dz .$$

Since the Casimir operator ω is bi-invariant, $E_m(z, gg_0)$ too is an eigenfunction of ω. Hence $T_{g_0}E_\psi \in \bigoplus_m L'_m$, by argument as in the preceding development. Thus $\bigoplus_m L'_m$ is an invariant subspace of the regular representation.

For a description of the equivalent representation of G on $\bigoplus_m \hat{L}'_m$, we introduce a holomorphic family of representations of G on H. For each z $\in \mathbf{C}$ we define $I(z, g_0) : H \to H$ by

$$(13) \qquad (I(z, g_0)h)(k(g)) = h(k(gg_0)) y(gg_0)^{\frac{(1+z)}{2}} y(g)^{\frac{-(1+z)}{2}} \qquad (h \in H) .$$

It is easily verified that the right hand side of the definition (13) depends on g only to the extent of its k-component, and thus indeed defines an element of H. In fact, $\{I(z) : \mathrm{Re}(z) = 0\}$ is equivalent to the principal series $\{\pi_z : \mathrm{Re}(z) = 0\}$ of irreducible unitary representations of $G = SL(2, \mathbf{R})$. As usual let $H(z)$ denote the set of all functions $\tilde{h}$ on G such that $\tilde{h}(nak) = y^{1/2}y^{z/2}h(k)$ for some $h \in H$, and $\tau_z : H \to H(z)$ be defined by $(\tau_z h)(nak) = y^{1/2}y^{z/2}h(k)$. Then it is easily checked that $I(z, g_0) = \tau_z^{-1} \circ T_{g_0} \circ \tau_z$. Of course, π_z is just the regular representation of G on $H(z)$, with $\mathrm{Re}(z) = 0$.

With the representations I(z) the action of G on the elements of L'_m can be expressed succinctly in terms of Eisenstein series. For $E_\psi \in L'_m$, we first have

$$T_{g_0}E_\psi(g) = \frac{1}{2\pi i} \int_{\mathrm{Re}(z)=0} E(z, \Psi(z), gg_0)\, dz .$$

If now we introduce the Eisenstein series

$$E(z, I(z, g_0)\Psi(z), g) = \sum_{\gamma \in \Gamma \cap P \backslash \Gamma} (I(z, g_0)\Psi(z))(\gamma g) y(\gamma g)^{\frac{(1+z)}{2}} \qquad (\mathrm{Re}(z) > 1)$$

then it follows immediately from the definition (13) that

$$E(z, I(z, g_0)\Psi(z), g) = E(z, \Psi(z), gg_0) .$$

Hence

$$T_{g_0}E_\psi(g) = \frac{1}{2\pi i}\int_{Re(z)=0} E(z, I(z, g_0)\Psi(z), g)\, dz . \tag{14}$$

More explicitly, we may write

$$I(z, g_0)e^{-im\theta} = \sum_{n\in 2Z} b_{mn}(z)e^{-in\theta} ,$$

where $b_{mn}(z)$ are entire functions. The right hand side of the equation (14) is then

$$\sum_n \frac{1}{2\pi i}\int_{Re(z)=0} b_{mn}(z)\hat{\psi}(z)E_n(z, g)\, dz .$$

By means of the Mellin inversion formula we can find smooth compactly supported functions ψ_{mn} $(n \in 2\mathbf{Z})$ on $N(\Gamma \cap P) \backslash G$ such that

$$\hat{\psi}_{mn}(z) = b_{mn}(z)\hat{\psi}(z) .$$

So the equation (14) becomes

$$T_{g_0}E_\psi = \sum_n E_{\psi_{mn}} .$$

We thus have a complete description of the equivalent representation of G on $\bigoplus_m \hat{\mathcal{L}}'_m$. For the preceding formula shows that $g_0 \in G$ sends

$$\Psi(z) + M_m(z)^{-1}\Psi(-z) \in \hat{\mathcal{L}}'_m$$

to

$$\sum_n [\, \Psi_{mn}(z) + M_n(z)^{-1}\Psi_{mn}(-z) \,] \in \bigoplus_{n\in 2Z} \hat{\mathcal{L}}'_n .$$

This completes our discussion of the spectral decomposition of the regular representation of G on $L^2(\Gamma \backslash G)$, to the extent of the introduction of Eisenstein series to analyze the continuous spectrum. We remark in closing that, using Eisenstein series, the unitary map between $\bigoplus_m L'_m$ and $\bigoplus_m \hat{L}'_m$ can be made explicit through rewritting the integral formula (12) as

$$E_\psi(g) = \frac{1}{2\pi i} \int_{Re(z)=0} E(z, F(z), g)\, dz$$

with

$$F(z) = \Psi(z) + M_m(z)^{-1}\Psi(-z) ,$$

where use has been made of the functional equations of $E(z, \Psi(z), g)$ and $M(z)$.

In the general case of a reductive real algebraic group G and an arithmetic subgroup Γ, the spectral decomposition of the regular representation, and so the introduction of Eisenstein series, requires a combinatorial framework based on the lattice structure of the collection of cuspidal subgroups of G. The following easily proven lemma (Harish-Chandra [2, p.4]) about cuspidal subgroups underlies this largely inductive development. Let (P, A) be a cuspidal pair of G, with the corresponding Langlands decomposition: P = NAM. The lemma asserts that if $\tilde{P}$ is a cuspidal subgroup of the Levi component M, with a Langlands decomposition: $\tilde{P} = \tilde{N}\tilde{A}\tilde{M}$, then there is a cuspidal subgroup P' of G contained in P which also has $\tilde{M}$ as a Levi component. One may in fact let $P' = N'A'M'$ where $M' = \tilde{M}$, $A' = \tilde{A}A$, and $N' = \tilde{N}N$. The correspondence $\tilde{P} \mapsto P'$ gives a bijection from the collection of cuspidal subgroups of M onto the collection of cuspidal subgroups of G contained in P.

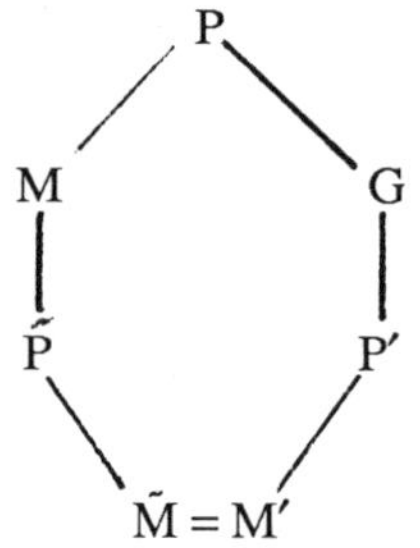

We give here some indication of how Eisenstein series generally arise in the spectral decomposition. One may refer to Langlands [1, 2, 3], Arthur [1], and Harish-Chandra [2] for in-depth treatment or details.

The collection of cuspidal subgroups of G is partitioned into finitely many equivalence classes of associate cuspidal subgroups. For each associate class we fix a complete set of representatives with respect to conjugation by elements of Γ. Since Γ-conjugation implies association, the union of all these sets is a complete set of representatives of the Γ-conjugacy classes of cuspidal subgroups. Specifically, these sets may be constructed as follows. Fix a minimal cuspidal subgroup P_0, and consider the standard cuspidal subgroups defined in relation to P_0. Partition the collection of all standard cuspidal subgroups of a given rank with respect to association. Suppose now $\{P_j\}$ is such an associate class of standard cuspidal subgroups. Let $\Xi \subset G_Q$ be consisted of one representative from each double coset in $\Gamma \backslash G_Q / (P_0)_Q$. For each j let $d(j) = |\Gamma \backslash G_Q / (P_j)_Q|$, and write Ξ as

$$\{\zeta_{11}, \ldots, \zeta_{1l_1}; \zeta_{21}, \ldots, \zeta_{2l_2}; \ldots; \zeta_{d(j)1}, \ldots, \zeta_{d(j)l_{d(j)}}\},$$

where ζ_{11} = the identity, and

$$\{\zeta_{11}, \ldots, \zeta_{1l_1}\}, \ \{\zeta_{21}, \ldots, \zeta_{2l_2}\} \ , \ldots, \ \{\zeta_{d(j)1}, \ldots, \zeta_{d(j)l_{d(j)}}\}$$

are the equivalence classes of $\Gamma \backslash \Xi / (P_j)_Q$. Introduce the index set

$$\underline{I}_j = \{11, 21, \ldots, d(j)1\}.$$

Set

$$\boldsymbol{P} = \bigcup_{j} \{ \zeta_{v_j} P_j \zeta_{v_j}^{-1} : v_j \in \underline{I_j} \} .$$

The collection $\boldsymbol{P}$ is then a complete set of representatives with respect to Γ-conjugation within an associate class. All associate classes are covered by this construction. In the special case of G = SL(2, **R**) there is but one standard cuspidal subgroup, namely P_0 itself, which is customarily chosen to be the subgroup of upper triangular matrices in G. If moreover it is assumed that $\Gamma \backslash G$ has only one cusp, then $\boldsymbol{P}$ consists just of P_0.

The spectral decomposition in general begins, as in the special case of G = SL(2, **R**), with the construction of generators for certain invariant subspaces of $L^2(\Gamma \backslash G)$ by means of the Eisenstein-Poincaré summation. Fix a maximal compact subgroup K. Let σ be an irreducible unitary representation of K on a finite dimensional inner product space V. A V-valued function f on G is called a σ-function if

$$f(gk) = \sigma(k)f(g) \qquad \text{for all } g \in G,\ k \in K .$$

For $P \in \boldsymbol{P}$, which has a natural Langlands decomposition: P = NAM, there are derived respectively from Γ and σ an arithmetic subgroup Γ_M of M and a representation σ_M of $K_M = K \cap M$ on V. Let ${}^0L^2(\Gamma_M \backslash M, \sigma_M, \chi)$ be the space of all V-valued cusp forms on M that are σ_M-functions, and are eigenfunctions for all of **Z**(M), the algebra of bi-invariant differential operators on M, for a given character χ of **Z**(M). If $\phi \in {}^0L^2(\Gamma_M \backslash M, \sigma_M, \chi)$ it may be extended to a σ-function on G by

$$\phi(g) = \phi(namk) = \sigma(k)\phi(m) \qquad (g = namk,\ n \in N,\ a \in A,\ m \in M,\ k \in K) .$$

Let ${}^0\underline{D}_{\sigma, P, \chi}$ be the span of V-valued functions on $N(\Gamma \cap P) \backslash G$ of the form $\psi(a)\phi(g)$ $(g = namk \in G)$, where $\phi(m)$ belongs to ${}^0L^2(\Gamma_M \backslash M, \sigma_M, \chi)$ and $\psi(a)$ is a smooth compactly supported complex-valued function on A. Finally, let ${}^0D_{\sigma, P, \chi}$ be the subspace of $C^\infty(N(\Gamma \cap P) \backslash G)$ spanned by functions of the form $\langle R(g), v \rangle_K$ $(g \in G)$, where $R \in {}^0\underline{D}_{\sigma, P, \chi}$ and

$v \in V$. The functions in ${}^0D_{\sigma, P, \chi}$ are compactly supported with respect to the a-component of g . For every $\psi \in {}^0D_{\sigma, P, \chi}$ we define

$$E_\psi(g) = \sum_{\gamma \in \Gamma \cap P \backslash \Gamma} \psi(\gamma g) \qquad (g \in G) .$$

Each E_ψ belongs to $L^2(\Gamma \backslash G)$. We denote by $L_{\sigma, \boldsymbol{P}}$ the closed subspace of $L^2(\Gamma \backslash G)$ generated by all such functions as P varies over $\boldsymbol{P}$ and, correspondingly, χ ranges over all characters of $\mathbf{Z}(M)$.

It can be established that $L^2(\Gamma \backslash G)$ is generated by the collection $\{L_{\sigma, \boldsymbol{P}}\}$ of all such subspaces:

$$L^2(\Gamma \backslash G) = \bigoplus_{\sigma, \boldsymbol{P}} L_{\sigma, \boldsymbol{P}} , \tag{15}$$

where $\boldsymbol{P}$ ranges over all sets of representatives with respect to Γ-conjugation within an associate class, and σ varies over a complete set of inequivalent irreducible unitary representations of K. We note first that if P and P' are Γ-congugates then, for a given σ, the span of all E_ψ constructed from P with a given character χ of $\mathbf{Z}(M)$ is equal to the span of all $E_{\psi'}$ constructed from P' with a corresponding character χ' of $\mathbf{Z}(M')$. On the other hand, for a given f(g) in $L^2(\Gamma \backslash G)$ of the form $\langle F(g), v \rangle$ with F being a square integrable σ-function on $\Gamma \backslash G$, and any $\psi(g)$ in ${}^0D_{\sigma, P, \chi}$ of the form $\langle \psi(a)\phi(g), v \rangle$ with $\phi(m)$ in ${}^0L^2(\Gamma_M \backslash M, \sigma_M, \chi)$, we have the following generalisation of the formula (6) for $\langle E_\psi , f \rangle$:

$$\langle E_\psi , f \rangle = \frac{\|v\|^2}{\dim V} \int_A \psi(a) e^{2\rho(\log a)} da \int_{\Gamma_M \backslash M} \langle \phi(m), F^P(am) \rangle \, dm , \tag{16}$$

where F^P is the constant term of F with respect to P, i.e., for P = NAM,

$$F^P(g) = \int_{N \cap \Gamma \backslash N} F(ng) \, dn ,$$

and ρ is one-half the sum of the positive roots of (P, A). (Note that we have, for simplicity, used the same symbol ψ for a compactly supported function on A as well as a function in ${}^0D_{\sigma, P, \chi}$.) If now $\langle E_\psi , f \rangle = 0$ for all E_ψ regardless of P, χ, ψ, ϕ, then it follows from the formula (16) that

(17) $$\int_{\Gamma_M \backslash M} \langle \phi(m), F^P(am) \rangle = 0 \qquad \text{for almost all a ,}$$

regardless of P, χ, ϕ. It is a theorem of Langlands that the condition (17) in turn implies that $F = 0$.

For P = G, whence M = G, the condition (17) in effect asserts that F is orthogonal to all V-valued cusp forms that are σ-functions, on account of the density of cuspidal automorphic forms. Also note the case of P being a minimal cuspidal subgroup. The condition simply expresses that $F^P = 0$. The proof of the theorem proceeds by induction on $\text{rank}_Q(G)$. In the induction step one shows that $F^P(am) = 0$ for all cuspidal subgroups different from G, i.e., F itself is a cusp form. This of course implies that F = 0. The proof of the vanishing of F^P for a proper cuspidal subgroup P appeals to the lemma on the lattice structure of cuspidal subgroups. Regard $F^P(am)$ as a function on M. Its constant term with respect to a cuspidal subgroup $\tilde{P}$ of M can be seen to equal $F^{P'}$, where P' is the corresponding cuspidal subgroup of G. As $\tilde{M} = M$, it follows that the condition (17) holds with the algebraic group M in place of G. Since $\text{rank}_Q(M) < \text{rank}_Q(G)$, $F^P(am) = 0$ by the induction hypothesis.

In view of standard harmonic analysis on the compact subgroup K, the completeness of the decomposition (15) is now quite clear. Regarding its orthogonality, with respect to $\boldsymbol{P}$ this is naturally understood in terms of Eisenstein series, whose primary role, as we saw in the SL(2, **R**) case, lies rather in the analysis of the continuous spectrum of the regular representation. For a cuspidal subgroup P in a particular $\boldsymbol{P}$ with the Langlands decomposition : P = NAM, we let $\boldsymbol{p}, \boldsymbol{a}, \boldsymbol{a}^*, \boldsymbol{a}_C, \boldsymbol{a}_C^*$ denote respectively the Lie algebra of P, the Lie algebra of A, the dual of $\boldsymbol{a}$, the complexified Lie algebra of A, and the dual of $\boldsymbol{a}_C$. Also, Σ^0 denotes the set of simple roots of $(\boldsymbol{p}, \boldsymbol{a})$. First, a smooth compactly supported function $\psi(a)$ on A can be expressed in terms of its Mellin transform $\hat{\psi}(\Lambda)$:

(18) $$\psi(a) = \left(\frac{1}{2\pi i}\right)^{\dim A} \int_{\mathrm{Re}(\Lambda)=\Lambda_0} \hat{\psi}(\Lambda) e^{(-\Lambda-\rho)(\log a)}\, d\Lambda \qquad (a \in A) ,$$

where

$$\hat{\psi}(a) = \left(\frac{1}{2\pi i}\right)^{\dim A} \int_A \psi(a) e^{(\Lambda+\rho)(\log a)}\, da \qquad (\Lambda \in \boldsymbol{a}_C^*)$$

and $\Lambda_0 \in \boldsymbol{a}^*$ is arbitrary. Let $H_{\sigma, P, \chi}$ be the subspace of $C^\infty(NA(\Gamma \cap P)\backslash G)$ spanned by functions of the form $\langle \phi(g), v \rangle$ for $\phi(m)$ in ${}^0L^2(\Gamma_M\backslash M, \sigma_M, \chi)$ and $v \in V$. We introduce the cuspidal Eisenstein series constructed with an element Φ of $H_{\sigma, P, \chi}$:

$$E(\Lambda, \Phi, g) = \sum_{\gamma \in \Gamma \cap P\backslash\Gamma} \Phi_\Lambda(\gamma g)$$

where

$$\Phi_\Lambda(g) = \Phi(g) e^{(-\Lambda-\rho)(\log a)} .$$

Using the reduction theory of arithmetic subgroups it can be established that $E(\Lambda, \Phi, g)$ converges absolutely for $\Lambda \in (\boldsymbol{a}_C^*)^+ = \{\Lambda \in \boldsymbol{a}_C^* : \langle \mathrm{Re}(\Lambda)-\rho, \alpha \rangle > 0 \text{ for all } \alpha \in \Sigma^o\}$. If now $\psi(g)$ is a function in ${}^oD_{\sigma, P, \chi}$ of the form $\psi(a)\Phi(g)$ with $\Phi \in H_{\sigma, P, \chi}$, we obtain from the expression (18) the following integral formula for E_ψ in terms of Eisenstein series:

$$(19) \qquad E_\psi(g) = \left(\frac{1}{2\pi i}\right)^{\dim A} \int_{\mathrm{Re}(\Lambda)=\Lambda_0} E(\Lambda, \Psi(\Lambda), g)\, d\Lambda ,$$

where

$$\Psi(\Lambda) = \Psi(\Lambda, g) = \hat{\psi}(\Lambda)\Phi(g) ,$$

$$E(\Lambda, \Psi(\Lambda), g) = \hat{\psi}(\Lambda) E(\Lambda, \Phi, g) ,$$

and $\Lambda_0 \in \boldsymbol{a}^*$ satisfies $\langle \Lambda_0-\rho, \alpha \rangle > 0$ for all $\alpha \in \Sigma^o$. There is then an inner product formula for E_ψ and $E_{\psi'}$, $\psi(g)$ and $\psi'(g)$ respectively in ${}^oD_{\sigma, P, \chi}$ and ${}^oD_{\sigma', P', \chi'}$, of respectively the forms $\psi(a)\Phi(g)$, $\psi'(a)\Phi'(g)$, with $\Phi \in H_{\sigma, P, \chi}$ and $\Phi' \in H_{\sigma', P', \chi'}$. Namely,

$$\langle E_\psi, E_{\psi'} \rangle = \left(\frac{1}{2\pi i}\right)^{\dim A} \int_{\mathrm{Re}(\Lambda)=\Lambda_0} \int_{A'} \int_K \int_{\Gamma_{M'}\backslash M'} E^{P'}(\Lambda, \Psi(\Lambda), a'm'k) \overline{\psi'(a'm'k)} e^{2\rho(\log a')}\, dm'dkda'd\Lambda .$$

Now a well-known computation of Langlands involving the Bruhat decomposition shows that

$E^{P'}(\Lambda, \Phi, g)$ is zero if P and P$'$ are not associate and rank P$'$ > rank P. The orthogonality of the decomposition (15) is then clear.

On the other hand, if P and P$'$ are associate to one another, the computation yields

$$E^{P'}(\Lambda, \Phi, g) = \sum_{s \in W(a, a')} e^{(-s\Lambda - \rho')(\log a')} M(s \mid \Lambda)\Phi(m'k) \qquad (g = n'a'm'k),$$

where $W(\boldsymbol{a}, \boldsymbol{a}')$ is the Weyl group of $(\boldsymbol{a}, \boldsymbol{a}')$; $M(s \mid \Lambda)$ is a linear transformation from $H_{\sigma, P, \chi}$ into $H_{\sigma, P', {}^s\chi}$, ${}^s\chi$ being the image of χ under the isomorphism induced by $s \in W(\boldsymbol{a}, \boldsymbol{a}')$ from $X(M)$, the set of all characters of $\mathbf{Z}(M)$, onto $X(M')$, the set of all characters of $\mathbf{Z}(M')$. Then

$$(20) \quad \langle E_\psi, E_{\psi'} \rangle = \left(\frac{1}{2\pi i}\right)^{\dim A} \int_{\mathrm{Re}(\Lambda)=\Lambda_0} \sum_{s \in W(a,a')} \left[\int_K \int_{\Gamma_{M'} \setminus M'} M(s \mid \Lambda)\Psi(\Lambda) \cdot \overline{\Psi'(-s\bar{\Lambda})}\, dm'dk \right] d\Lambda .$$

This inner product formula suggests that each subspace $L_{\sigma, \boldsymbol{P}}$ in the decomposition (15) can itself be decomposed. For $P \in \boldsymbol{P}$, the action of the Weyl group of $\boldsymbol{a}$ on $X(M)$ partitions $X(M)$ into disjoint orbits. It can be shown that if χ_1 and χ_2 belong to distinct orbits in $X(M)$, then arbitrary Φ_1 and Φ_2 from H_{σ, P, χ_1} and H_{σ, P, χ_2} respectively are orthogonal. If ξ is an orbit in $X(M)$, and P$'$ is an associate of P, on account of the Weyl group $W(\boldsymbol{a}, \boldsymbol{a}')$ there is a corresponding orbit ξ' in $X(M')$. Let $\{\xi\}$ denote a collection of such corresponding orbits as P$'$ varies over $\boldsymbol{P}$. Define $L_{\sigma, \boldsymbol{P}, \{\xi\}}$ to be the closed subspace of $L_{\sigma, \boldsymbol{P}}$ spanned by all E_ψ as P$'$ varies over $\boldsymbol{P}$ and, correspondingly, χ ranges over $\xi' \in \{\xi\}$. For simplicity we index the set of all triples $(\sigma, \boldsymbol{P}, \{\xi\})$ by κ. In place of the decomposition (15) we can now state the primary decomposition of $L^2(\Gamma \setminus G)$:

$$(21) \qquad L^2(\Gamma \setminus G) = \bigoplus_\kappa L_\kappa .$$

The combinatorial framework we have set up in the course of establishing the primary decomposition naturally underlies the theory of Eisenstein series as well. Suppose for the moment that we have the meromorphic continuation and functional equations of all cuspidal

Eisenstein series, and of all $M(s \mid \Lambda)$, then we can begin to analyze each subspace L_{κ} by means of the cuspidal Eisenstein series for the cuspidal subgroups in $\boldsymbol{P}$. If E_{ψ} belongs to an appropriate closed subspace $L_{\kappa, \boldsymbol{P}}$ of L_{κ}, $L_{\kappa, \boldsymbol{P}}$ being the generalization of L_m' in the SL(2, **R**) case, the contour of integration in the formulae (19) and (20) may be shifted to $\mathrm{Re}(\Lambda) = 0$. Thus in the beginning stage the analysis of L_{κ} proceeds much as that of L_m in the SL(2, **R**) case.

In general however, the orthogonal complement $L_{\kappa, \boldsymbol{P}}{}^{\perp}$ of $L_{\kappa, \boldsymbol{P}}$ in L_{κ} still lends itself to be analyzed by means of Eisenstein series. By the residue theorem the elements of $L_{\kappa, \boldsymbol{P}}{}^{\perp}$ are at first rank $\boldsymbol{P} - 1$ dimensional integrals of residues of Eisenstein series. But these residues can be interpreted as *noncuspidal* Eisenstein series for larger cuspidal subgroups. We give a rough indication of the idea. Suppose $E(\Lambda, \Psi(\Lambda), g)$, which equals $\hat{\psi}(\Lambda)E(\Lambda, \Phi, g)$ for $\Phi \in H_{\sigma, P, \chi}$, is a cuspidal Eisenstein series for P in $\boldsymbol{P}$. Its singularities in the region extending from the domain of absolute convergence up to $\mathrm{Re}(\Lambda) = 0$ lie on hyperplanes of the form $\boldsymbol{r} = X(\boldsymbol{r}) + \boldsymbol{r}^{o}_{C}$, where $\boldsymbol{r}^{o}$ is a subspace of $\boldsymbol{a}^*$ of codimension one, and $X(\boldsymbol{r})$ is a vector in $\boldsymbol{a}^*$ orthogonal to $\boldsymbol{r}^{o}$. It may be supposed that $\boldsymbol{r}^{o}$ equals $(\boldsymbol{a}')^*$ for some cuspidal subgroup P' containing P, with $P' \in \boldsymbol{P}'$. We now recall the lemma on the lattice structure of cuspidal subgroups, renotated for present use:

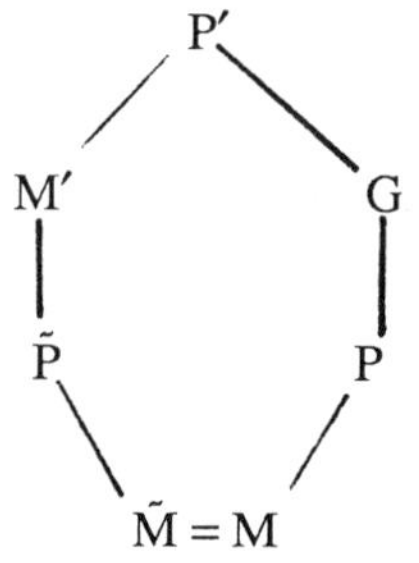

which asserts the existence of $\tilde{P}$ and $\tilde{M}$. As $\tilde{M} = M$, $\Phi(m)$ may also be regarded as a cuspidal automorphic form on $\tilde{M}$. On the other hand, the Eisenstein series $E(\Lambda, \Psi(\Lambda), g)$ can be decomposed as follows:

$$E(\Lambda, \Psi(\Lambda), g) = \sum_{\gamma \in \Gamma \cap P \backslash \Gamma} \Psi(\Lambda, \gamma g) e^{(-\Lambda-\rho)(\log a(\gamma g))}$$

$$= \sum_{\gamma \in \Gamma \cap P' \backslash \Gamma} \sum_{\delta \in \Gamma \cap P \backslash \Gamma \cap P'} \Psi(\Lambda, \delta\gamma g) e^{(-\Lambda-\rho)(\log a(\delta\gamma g))}$$

$$(22) \qquad = \sum_{\gamma \in \Gamma \cap P' \backslash \Gamma} E(\tilde{\Lambda}, \Psi(\Lambda), m'(\gamma g)) e^{(-\Lambda'-\rho)(\log a'(\gamma g))}$$

for $\Lambda = \Lambda' + \tilde{\Lambda}$ where $\tilde{\Lambda} \in (\tilde{a})^*_C$, and $\Lambda' \in (\boldsymbol{a}')^*_C$ is such that the last sum converges. The function $E(\tilde{\Lambda}, \Psi(\Lambda), m'(g))$ is a cuspidal Eisenstein series on the algebraic group M' for the rank one cuspidal subgroup $\tilde{P}$, and is thus a meromorphic function on the whole of $(\tilde{a})^*_C$. Moreover, its residue $\Theta(\Lambda')$ at $X(\boldsymbol{r})$ for a singular hyperplane $\boldsymbol{r}$ is easily seen to be a square integrable automorphic form on M' (multiplied with an entire function of Λ'), which is however noncuspidal. Now

$$\mathrm{Res}_{\boldsymbol{r}} E(\Lambda, \Psi(\Lambda), g) = \frac{1}{2\pi i} \int_0^{2\pi} E(\Lambda + \tilde{\Lambda}_\theta, \Psi(\Lambda), g)\, d\theta$$

$$= \frac{1}{2\pi i} \int_0^{2\pi} E(\Lambda' + X(\boldsymbol{r}) + \tilde{\Lambda}_\theta, \Psi(\Lambda' + X(\boldsymbol{r})), g)\, d\theta$$

where $\tilde{\Lambda}_\theta = \varepsilon e^{2\pi i\theta} X(\boldsymbol{r}) \in (\tilde{a})^*_C$, with a small positive number ε. Therefore the residue of the original Eisenstein series $E(\Lambda, \Psi(\Lambda), g)$ at $\boldsymbol{r}$, parametrized by $\Lambda' \in (\boldsymbol{a}')^*_C$, is an Eisenstein series $E(\Lambda', \Theta(\Lambda'), g)$ for the cuspidal subgroup P'.

Being a residue of a cuspidal Eisenstein series, the noncuspidal Eisenstein series $E(\Lambda', \Theta(\Lambda'), g)$ clearly has meromorphic continuation to the whole of $(\boldsymbol{a}')^*_C$. Furthermore, by the same token, its constant term with respect to a cuspidal subgroup in $\boldsymbol{P}'$ can be expressed in terms of linear transformations $M(s' \mid \Lambda')$ which too have meromorphic continuation. Using these the analysis of L_K may now proceed further, with the introduction of a subspace $L_{K, \boldsymbol{P}'}$ of $L_{K, \boldsymbol{P}}{}^{\perp}$, in analogy with $L_{K, \boldsymbol{P}}$ in L_K. Note that rank $\boldsymbol{P}'$ = rank $\boldsymbol{P} - 1$. Thus it is that an inductive development yields the complete analysis of L_K by Eisenstein series.

It should be stressed that in the preceding development of spectral decomposition one needs to assume that the Eisenstein series $E(\tilde{\Lambda}, \Psi(\Lambda), m'(g))$ on M' and its like in the later stages all have meromorphic continuation. This is clear only in the second stage of the development, when $E(\tilde{\Lambda}, \Psi(\Lambda), m'(g))$ is derived from a cuspidal Eisenstein series and therefore is itself a cuspidal Eisenstein series. At every later stage there will however be a noncuspidal Eisenstein series for a maximal cuspidal subgroup of a lower rank algebraic group. The difficulty resolves if it is known already that Eisenstein series on algebraic groups of ranks lower than that of G all have meromorphic continuation. On the other hand, for an arbitrary cuspidal subgroup P in some ***P***, a noncuspidal Eisenstein series for P is formed with an element in the discrete spectrum of its Levi component M. If one makes use of the spectral decompositions of $L^2(\Gamma' \setminus G')$ for algebraic groups G' of ranks lower than that of G, one can show that an arbitrary noncuspidal Eisenstein series on G is necessarily a several-fold residue of a cuspidal Eisenstein series on G. Therefore, the analysis of the subspace L_K should be incorporated into another, larger, induction on $\mathrm{rank}_Q G$, in which one establishes the meromorphic continuation and functional equations of all Eisenstein series as well as the spectral decomposition. The meromorphic continuation of noncuspidal Eisenstein series, which arise naturally in the context of spectral theory, is thus interwoven into the development of spectral decomposition.

We come now to consider the meromorphic continuation and functional equations of cuspidal Eisenstein series, which underlay all of the foregoing discussion. In Langlands' theory, one first establishes for the case of Eisenstein series for rank one cuspidal subgroups, from which the general case in then deduced. The case of maximal cuspidal subgroups is fundamental because here it is utilized much of the insight into the analytic nature of Eisenstein series as needed for their meromorphic continuation. For this Langlands was inspired by a very brief sketch of Selberg [3]. A key ingredient in obtaining the general case from the fundamental case is the decomposition of an arbitrary cuspidal Eisenstein series in terms of a cuspidal Eisenstein series for a maximal cuspidal subgroup, as in the decomposition (22) above. This, as we saw, is a manifestation of the lattice structure of cuspidal subgroups.

In this work we shall develop another approach of Selberg, one which makes use of the Fredholm theory of integral equations, to obtain the meromorphic continuation and functional equations of cuspidal Eisenstein series for the maximal cuspidal subgroups of a reductive real algebraic group G, with an arithmetic subgroup Γ. The groups G and Γ are basically of the types considered by Harish-Chandra in [2], except that in the course of establishing Selberg's eigenfunction principle we shall have to restrict G somewhat. This may however be an unnecessary complication; it is likely that there are alternatives to our derivation of the Selberg principle.

In order that the analytic development may not be obscured by the elaborate formalism of the general combinatorial situation, we have assumed in the first five chapters that $\Gamma \backslash G$ has only one cusp. The general case of several cusps will be presented in the last chapter. Denote by $E(\Lambda,g)$, with $g \in G$ and $\Lambda \in (\mathfrak{a}_C^*)^+$, the Eisenstein series constructed with a cusp form on the Levi component M of a rank one standard cuspidal subgroup P. A basic property of $E(\Lambda, g)$ that follows readily from its definition is that it is an eigenfunction for all the bi-invariant differential operators on G. Let K be a fixed maximal compact subgroup of G, and σ be a character of K. The starting point of this approach to the meromorphic continuation of Eisenstein series is the well known principle of Selberg which says in this context that a σ function on G that is an eigenfunction for all the invariant differential operators must also be an eigenfunction for all the convolution operators (Chapter 1). One can easily construct a compactly supported kernel function α so that

$$\alpha * E(\Lambda, g) = \int_G \alpha(h^{-1}g)E(\Lambda, h)\, dh$$

$$= \hat{\alpha}^o(\Lambda)E(\Lambda, g) \qquad (g \in G, \Lambda \in (\mathfrak{a}_C^*)^+),$$

where $\hat{\alpha}^o$ is a Selberg transform, which can be shown to be a nonconstant entire function on $\mathfrak{a}_C^*$.

Rewrite this integral equation as

(23) $$(\alpha * - \hat{\alpha}^{o}(\Lambda))E(\Lambda, g) = 0 .$$

We shall use the general results of operator theory, instead of appealing specifically to the theory of integral equations. The resolvent of an operator is holomorphic everywhere except on the spectrum of the operator. The idea then, roughly speaking, is to truncate $\alpha *$ down to a compact operator $\mathbf{K}_{\alpha} : L^2(\Gamma \backslash G) \to L^2(\Gamma \backslash G)$ so that spec $\mathbf{K}_{\alpha}$ is suitably sparse.

The analysis needed for the truncation is adapted from Godement's exposition ([2]) of Langlands' theory. Essentially one subtracts from the kernel $\alpha(h^{-1}g)$ each of its constant terms with respect to rank one standard cuspidal subgroups, and obtains estimates over a Siegel domain associated with the fixed minimal cuspidal subgroup. The Siegel domain is supposed to be large enough so that it contains a fundamental domain F for $\Gamma \backslash G$. On the other hand, for reasons that will be apparent later (Chapter 4), we require that $\mathbf{K}_{\alpha}$ acts smoothly. To ensure this, a smoothing operator λ_{α} is applied on top of a truncated operator $\mathbf{K}_{\alpha}^{o}$; together they form $\mathbf{K}_{\alpha}$ (Chapter 2). It will be necessary to introduce certain truncation factors δ_j .

Upon truncation, the equation (23) becomes

(24) $$(\mathbf{K}_{\alpha} - \hat{\alpha}(\Lambda))E(\Lambda, g) = -\hat{\alpha}^{o}(\Lambda) \sum_{j} \lambda_{\alpha}[\, \delta_j(g)E^j(\Lambda, g)\,]_F$$

where $E^j(\Lambda, g)$ is the constant term of $E(\Lambda, g)$ with respect to the jth rank one standard cuspidal subgroup, $[\ \ ]_F$ denotes automorphic extension from F to all of G, and $\hat{\alpha} = (\hat{\alpha}^{o})^3$ (Chapter 3). Now, since P is itself of rank one, for each j Langlands' formula for the constant terms of cuspidal Eisenstein series enables us to express each of the summands in the equation (24) as

$$\sum_{s,n} \phi(j, s, n \mid \Lambda) H^{o}_{j,s,n}(\Lambda, g) ,$$

where the $H^{o}_{j,s,n}(\Lambda, g)$ are holomorphic in Λ for $\Lambda \in \boldsymbol{a}_{C}^{*}$, and the constant term coefficients $\phi(j, s, n \mid \Lambda)$ are derived from intertwining operators $M(s \mid \Lambda)$, which are holomorphic on $(\boldsymbol{a}_{C}^{*})^{+}$, where the Eisenstein series was originally defined. For simplicity, use a multiindex J in place of (j, s, n). We then have

$$(25)\qquad (\mathbf{K}_\alpha - \hat{\alpha}(\Lambda))E(\Lambda, g) = -\sum_J \phi(J \mid \Lambda)\hat{\alpha}(\Lambda)H_J(\Lambda, g) \qquad H_j = \frac{H_J^o}{(\hat{\alpha}^o)^2} .$$

This last equation suggests that we solve the system of equations:

$$(\mathbf{K}_\alpha - \hat{\alpha}(\Lambda))F_J^{**}(\Lambda, g) = -\mathbf{K}_\alpha H_J(\Lambda, g)$$

for F_J^{**} in $L^2(\Gamma \setminus G)$, and define new functions:

$$F_J^*(\Lambda, g) = F_J^{**}(\Lambda, g) + H_J(\Lambda, g) ;$$

so that

$$F = \sum_J \phi(J)F_J^*$$

satisfies the equation (25) like E does. Using estimates analogous to those in the truncation of the integral operator, it can be established that the function

$$E(\Lambda, g) - \sum_J \phi(J \mid \Lambda)H_J(\Lambda, g) ,$$

as well as

$$F(\Lambda, g) - \sum_J \phi(J \mid \Lambda)H_J(\Lambda, g) ,$$

belongs to $L^2(\Gamma \setminus G)$. (We call this the truncation of the Eisenstein series.) It then follows that

$$(26)\qquad E(\Lambda, g) = \sum_J \phi(J \mid \Lambda)F_J^*(\Gamma, g) .$$

As the F_J^* are by construction already holomorphic on $\boldsymbol{a}_C^* \setminus \hat{\alpha}^{-1}(\text{spec } \mathbf{K}_\alpha)$ for each $g \in G$, the equation (26) reduces the problem of analytic continuation of $E(\Lambda, g)$ to that of $\phi(J \mid \Lambda)$.

In continuing $\phi(J \mid \Lambda)$ (Chapter 4) direct use will be made of the fact that $E(\Lambda, g)$ is an eigenfunction of the invariant differential operators:

$$DE = \chi_\Lambda(D)E, \qquad D \in \mathbf{Z}(G).$$

Thus, using the equation (26), we get

$$\sum_J \phi(J \mid \Lambda)\,[\,DF_J^* - \chi_\Lambda(D)F_J^*\,] = 0, \qquad D \in \mathbf{Z}(G).$$

To these basic equations we add some others also based on the known properties of the Eisenstein series (as a function of g) and thus obtain a system of linear equations with coefficients which are everywhere meromorphic, and with $\phi(J \mid \Lambda)$ as unknowns. We will solve this system for $\phi(J \mid \Lambda)$ uniquely, and have their full meromorphicity as a consequence. This system of equations refers to a function $\sum_J \phi(J \mid \Lambda)F_J^*(\Lambda, g)$ possessing some properties (in g) which go back to the Eisenstein series itself. Now nonuniqueness implies the existence of two different functions sharing certain properties; one of these is the growth of constant terms. But, on the other hand, it can be shown, using the self-adjointness of the Casimir operator in $\mathbf{Z}(G)$. that two functions possessing such properties must be identical. Thus we arrive at a contradiction. It should be mentioned that because we are dealing with a cuspidal Eisenstein series for a maximal standard cuspidal subgroup, the growth of its constant terms is especially transparent; the number of associate maximal standard cuspidal subgroups is either 1 or 2, and the respective Weyl groups appearing in Langlands' formula for constant terms are particularly simple.

Along this line of thought we can interpret the functional equation of the Eisenstein series as a uniqueness statement. We will verify that both sides of the functional equation possess the same set of characterizing properties as above (Chapter 5).

Selberg's earlier approach, on which Langlands' treatment of the fundamental case was based, proceeds with a somewhat different emphasis. This can be quite adequately illustrated with the $SL(2, \mathbf{R})$ case, and specifically the right K-invariant Eisenstein series $E_0(z, g)$ ($m = 0$), which may be regarded as defined on the upper half-plane $\mathbf{H}$. For simplicity we assume that

$\Gamma\backslash\mathbf{H}$ has only one cusp, and which is at ∞. Also, in place of $E_0(z, g)$ we write $E(z, \tau)$ ($\tau \in \mathbf{H}$). We indicate briefly the main points of this development, for a detailed treatment one may refer to Kubota [1]. The constant term of $E(z, \tau)$ is

$$y^{\frac{(1+z)}{2}} + M(z)y^{\frac{(1-z)}{2}} \qquad (\tau = x + iy \in \mathbf{H}) .$$

In this approach one investigates primarily $M(z)$, and deduces from its analytic properties those of the Eisenstein series $E(z, \tau)$. Use is made of two inner product formulae in which $M(z)$ appears. The first of these is the inner product formula (10). The second inner product formula is the Maass-Selberg relation, which results from an application of Green's theorem to Eisenstein series over a truncated fundamental domain.

The analytic continuation of $M(z)$, and thence of $E(z, \tau)$, proceeds in two steps. The first step is to continue into the region $\{z : \mathrm{Re}(z) > 0\} \setminus (0, 1]$. Let $\hat{\psi}(z)$ be the Mellin transform of a smooth compactly supported function $\psi(y)$. We begin with the function

$$\psi_\zeta(y) = \frac{1}{2\pi i} \int_{\mathrm{Re}(z)=z_0} \frac{\hat{\psi}(z)}{(1-z^2)-(1-\zeta^2)}\, y^{\frac{(1+z)}{2}}\, dz .$$

Provided $\mathrm{Re}(\zeta) > z_0$, $\psi_\zeta(y)$ is bounded as $y \to \infty$. If $z_0 > 1$ as well, then $E_{\psi_\zeta}(\tau)$ is well defined and belongs to $L^2(\Gamma\backslash\mathbf{H})$. Moreover,

$$E_{\psi_\zeta}(\tau) = \frac{1}{2\pi i} \int_{\mathrm{Re}(z)=z_0} \frac{\hat{\psi}(z)}{(1-z^2)-(1-\zeta^2)}\, E(z, \tau)\, dz .$$

It follows from the eigenfunction property of the Eisenstein series that

$$(D-(1-\zeta^2))E_{\psi_\zeta} = E_\psi ,$$

if the Laplace-Beltrami operator D is taken to be $-4y^2\left(\frac{\partial^2}{\partial x^2}+\frac{\partial^2}{\partial y^2}\right)$. Then

(23) $$E_{\psi_\zeta} = R(1-\zeta^2)E_\psi \qquad (1-\zeta^2 \notin \mathrm{spec}\, D) ,$$

where $R(1-\zeta^2)$ is the resolvent of D. Note that spec D is real and nonnegative. On the other hand, if the contour of integration in the inner product formula (10), as applied to $\langle E_{\psi_\zeta}, E_\psi \rangle$, is shifted to the right of ζ, then an application of the residue theorem yields a formula for $M(\zeta)$. This together with the formula (23) enables us to continue $M(\zeta)$ to the said region. The analytic continuation of the Eisenstein series is then accomplished with power series expansions whose centers move into this region from the domain of absolute convergence. On the basis of the analytic continuation of $M(\zeta)$, the Maass-Selberg relation furnishes the needed estimates for the radii of convergence. In the second step one shows that limit exists for M(z) as z approaches the remaining part of the right half-plane $\mathrm{Re}(z) \geq 0$ (∞ is allowed as a limit), and also establishes its functional equation on the line $\mathrm{Re}(z) = 0$. The meromorphic continuation of M(z) to the whole complex plane then follows immediately. The main tool in the second step is the Maass-Selberg relation, through which one exploits the symmetry of the constant term of the Eisenstein series. Rather delicate functional analytic arguments are needed for this, as well as to deduce the corresponding properties of the Eisenstein series itself.

Closer in spirit to our approach is that of Roelcke [2]. There direct use is made of the analytic properties of an Eisenstein series $E(z, \tau)$ as function of τ to effect its meromorphic continuation in z. As before we assume that $\Gamma \setminus \mathbf{H}$ has only one cusp, at ∞. Let $\Phi(z, \tau)$ be a smooth automorphic function that equals $y^{(1+z)/2}$ for all sufficiently large y. One works with the truncated Eisenstein series $u(z, \tau) = E(z, \tau) - \Phi(z, \tau)$. For $\mathrm{Re}(z) > 0$, $u(z, \tau)$ belongs to $L^2(\Gamma \setminus \mathbf{H})$. Now the eigenfunction property of the Eisenstein series implies that

(24) $$(D - (1 - z^2))u = \Psi(z, \tau),$$

where

$$\Psi(z, \tau) = (D - (1 - z^2))\Phi,$$

and $\Psi(z, \tau)$ is also in $L^2(\Gamma \setminus \mathbf{H})$. One then appeals to the spectral theory of the Laplace-Beltrami operator to continue $u(z, \tau)$, and thence $E(z, \tau)$. By its very nature this approach is limited by one's knowledge about the spectrum of D in $L^2(\Gamma \setminus \mathbf{H})$. It is perhaps worth noting that Roelcke, for rather technical reasons, actually convert the equation (24) to an integral equation through the use of a Green's function, before effecting the analytic continuation. However, without the benefit of the more flexible Selberg's eigenfunction principle, the method as it stood was not enough to fully establish the meromorphic continuation of the Eisenstein series.

It is the common merit of such broad, functional analytic approaches that the meromorphic continuation and functional equations of Eisenstein series are seen to be consequences of the basic analytic properties they possess by virtue of their construction, and do not depend on the specific arithmetic nature of the discrete subgroups. Another interesting example of this is the approach of P. Lax and R. Phillips, which uses notions of scattering theory (Lax and Phillips [1]). On the other hand, these approaches do not lend themselves readily to deeper investigation of the meromorphicity of Eisenstein series, especially regarding the location of their poles. It is of obvious interest to incorporate arithmetic information about the discrete subgroups into one of these general developments. This seems to be a very difficult matter, however.

Noteworthy perhaps in our approach is that the Eisenstein series can be expressed completely in terms of the Fredholm solutions. Whether such expressions are useful for a deeper understanding of the Eisenstein series is uncertain, as our knowledge of the Fredholm solutions themselves is still very limited. A matter of special interest in the whole development is the choice of α used in the construction of the compact operator $\mathbf{K}_\alpha$; there seems to be a lot of freedom in this. In fact, for a given choice of α , each of the functions $F_J^*(\Lambda, g)$ appearing in the equation (26) is only guaranteed by the spectral theory of compact operators to be holomorphic on $\boldsymbol{a}_{\mathbf{C}}^* \setminus \hat{\alpha}^{-1}(\text{spec } \mathbf{K}_\alpha)$ and has a pole at each point of $\hat{\alpha}^{-1}(\text{spec } \mathbf{K}_\alpha \setminus \{0\})$. These poles may very well accumulate at $\hat{\alpha}^{-1}(\{0\})$. It is only in conjunction with another choice of kernel function that the meromorphic continuation of $E(\Lambda, g)$ will be fully established. Naturally we would like to clarify the role of the kernel function α in this approach to the meromorphic continuation of Eisenstein series. It is conceivable that some judicious choices of kernel in specific cases of G and Γ will lead to quite precise information about the poles of the Eisenstein series. Indeed, circumstances at the time of this writing prevented us from carrying this development beyond the bare minimum of meromorphic continuation and functional equations of cuspidal Eisenstein series for maximal cuspidal subgroups. Certain facts about their poles will not be derived here. For instance, it is known that the set of poles of the Eisenstein series is contained in the set of poles of the intertwining operators, about which there is some general information (Harish-Chandra [2, p. 105]). But we think that these general assertions are well within the framework of our development, and can be obtained without undue effort.

That analysis of an Eisenstein series $E(\Lambda, g)$ as a function of the group variable g should lead to information about it as a function of Λ is all the more remarkable when one considers the early, isolated occurrences of Eisenstein series in analytic number theory. Apart from their central importance in the theory of automorphic forms as we saw in the above, Eisenstein series have other, more immediate, applications to analytic number theory. For an arithmetically defined Γ, the Eisenstein-Poincaré summation encodes into the Eisenstein series arithmetic information much in the same way the construction of a Dirichlet series or zeta function does. Thus it is not altogether unexpected that specialization in the group variable may yield connections with Dirichlet series or zeta functions. A well known early instance of this is Dirichlet's relation between zeta functions of imaginary quadratic number fields and special values of the Eisenstein series for the modular group, which he derived in the course of proving his class number formula (Goldfeld [1]). Another, more far reaching connection of a similar kind lies in the constant terms of cuspidal Eisenstein series. These are known to involve simple combinations of zeta functions or L-functions. Clearly, in this regard knowledge of the meromorphicity of Eisenstein series is highly desirable. As an example, Shahidi (Gelbart and Shahidi, [1]) has applied Langlands' theory of Eisenstein series to prove Langlands' conjecture on his Euler product global L-functions in a number of special cases. Of course, Eisenstein series have also appeared in the guise of Epstein zeta functions, which found quite many uses in analytic number theory (Terras [1], Siegel [1]).

The conceptual simplicity of the present approach invites one to attempt the development with cuspidal Eisenstein series for higher rank cuspidal subgroups. On the other hand, treatment of noncuspidal Eisenstein series by this method seems to pose greater challenge. A formula like Langlands' for constant terms of cuspidal Eisenstein series, which makes transparent their growth with respect to the group variable, has first to be worked out. Recall that in Langlands' theory noncuspidal Eisenstein series are treated by means inseperable from the spectral decomposition. Presently no other general approach for noncuspidal Eisenstein series is known (Langlands [2]). In any of these potential developments of our method, possibly towards providing an alternative to Langlands' theory, one is likely to need more elaborate uniquness theorems than our Theorem 4.3.1. This will depend on the shape of the constant term, and may occasion some rather interesting analyses.

The present work is based on a rough sketch given by Professor Selberg to Professors C. Moreno and A. Terras at the Institute for Advanced Study in March of 1984. During its writing I also found a write-up of P. Cohen and P. Sarnak (Cohen and Sarnak [1]) for the SL(2, **R**) case to be quite instructive. (Incidentally, this case had also been worked out in Hejhal [1].) Apart from some typographical revisions, the body of this work is the doctoral thesis I submitted to the University of Illinois at Urbana-Champaign in September of 1987. It contains several glaring inadequacies. Some are of a technical nature, others are omissions forced upon by lack of time. Yet a number of people have favored its publication, on the ground that Selberg's new approach should be made better known to the mathematical community. Let me record my appreciation for the encouragement of Professors S. Friedberg and A. Terras. I am also grateful to Professor

R. Rao for pointing out an error in Chapter 1 and for generously providing the correct argument. Finally, I thank my teacher Professor Carlos Moreno for his instruction and guidance.

CHAPTER 1

THE DEFINITION AND BASIC PROPERTIES OF THE EISENSTEIN SERIES

1.1. Introduction

We begin this development by presenting background material necessary for the definition of the Eisenstein series. We then define the Eisenstein series and list some of its basic properties which are relevant for the meromorphic continuation. All the material in this chapter is standard and, with the possible exception of Selberg's eigenfunction principle, good references are available. Here, as in other parts of this work, the author draws heavily from the works of Harish-Chandra, especially his monograph: *Automorphic Forms on Semi-simple Lie Groups.* This book begins with a presentation of some basic results from the structure theory of reductive real algebraic groups. In particular, basic facts about cuspidal subgroups are stated. Many of these essentially came from the theory of parabolic subgroups of real semi–simple Lie groups. Schlichtkrull [1] has a summary of the latter theory; for more details one can refer to Varadarajan [2] and Warner [1].

As to Selberg's principle, we have made reference to Selberg's original exposition [2]. For our development we do not need the principle in its full generality, but only the ability to regard an eigenfunction for the bi-invariant differential operators on a real reductive Lie group $\tilde{G}$ as an eigenfunction for convolution operators.

Denote the set of all bi-invariant differential operators by $\mathbf{Z}(\tilde{G})$, and the set of all left $\tilde{G}$-invariant differential operators by $\boldsymbol{D}(\tilde{G})$. Presently let K denote a maximal compact subgroup of $\tilde{G}$. The idea behind our exposition is that any smooth function on $\tilde{G}$ which is $\mathbf{Z}(\tilde{G})$-finite and right K-finite is real-analytic, and that two analytic functions f_1 and f_2 are equal if at some point $g \in \tilde{G}$,

$$(Df_1)(g) = (Df_2)(g) \quad \text{for all } D \in \boldsymbol{D}(\tilde{G}) .$$

Let σ be a character of K, and Φ be a σ-function on $\tilde{G}$ that is an eigenfunction for all of $\mathbf{Z}(\tilde{G})$. The first important step is to show that Φ is actually an eigenfunction for all of $\boldsymbol{D}_K(\tilde{G})$, the left $\tilde{G}$-invariant and right K-invariant differential operators (Theorem 1.4.2). Since we do not require that $\tilde{G}$ be connected, certain assumptions (Harish-Chandra [3, pp. 105-106]; Schlichtkrull [1, pp. 56-57, 59-61]) need to be made in order that the standard results about the algebra of invariant differential operators of a connected semi-simple Lie group be still valid; these simplify somewhat in the setting of algebraic groups. We shall need a result on the structure of the universal enveloping algebra of $\tilde{G}$ (Theorem 1.4.1), which is valid under the additional assumption that the Lie algebra of $\tilde{G}$ is a real classical semi-simple Lie algebra. Passing from $\boldsymbol{D}_K(\tilde{G})$ to $\boldsymbol{D}(\tilde{G})$ is however a less obvious matter. Following Selberg, we shall introduce the symmetrization operator

$$M_z : C^\infty(\tilde{G}) \to C^\infty(\tilde{G}) : \Phi \mapsto M_z\Phi .$$

The crucial properties of M_z are:

1) (Proposition 1.4.2) For any $D \in \boldsymbol{D}(\tilde{G})$,

$$D(M_z\Phi)(z) = D_K(M_z\Phi)(z) ,$$

where

$$D_K = \int_K \mathrm{Ad}(k)D\, dk \in \boldsymbol{D}_K(\tilde{G}) ;$$

2) $D(M_z\Phi) = M_z D\Phi$ for all $D \in \boldsymbol{D}(\tilde{G})$;

3) $(M_z\Phi)(z) = \Phi(z)$ if Φ is a smooth σ-function;

4) $\alpha * M_z\Phi = M_z(\alpha * \Phi)$;

where α is a continuous compactly supported function on $\tilde{G}$, and the convolution $\alpha * \Phi$ is defined by

$$(\alpha * \Phi)(g) = \int_G \alpha(g'^{-1}g)\Phi(g')\,dg' \qquad (g \in \tilde{G}) .$$

The first significant conclusion will actually be a corollary to a result on the application of the symmetrization operator M_Z to Φ and to $\alpha * \Phi$ (Corollary to Theorem 1.4.3).

Lastly, the Selberg transform $\hat{\alpha}^o(\Lambda)$ that appears in the case of Selberg's principle in which we are interested will be shown to be a nonconstant entire function, provided that it is not identically zero (Proposition 1.4.4). The significance of $\hat{\alpha}^o(\Lambda)$ will be clear by Chapter 3 when we introduce the Fredholm equations.

1.2. Prerequisites for the Eisenstein Series

In this work, as in Harish-Chandra [2], $G \subset GL(n, \mathbf{R})$ is a *reductive real algebraic group* of the following type: Let $\underline{\underline{G}} \subset GL(n, \mathbf{C})$ be a reductive algebraic group defined over $\mathbf{Q}$ satisfying

1) $\chi^2 = 1$ for any $\chi \in X_{\mathbf{Q}}(\underline{\underline{G}})$, which is the group of rational characters of $\underline{\underline{G}}$ define over $\mathbf{Q}$,

2) if $\underline{\underline{T}}$ is a maximal torus of $\underline{\underline{G}}$, then $Z(\underline{\underline{T}})$ meets every connected component of $\underline{\underline{G}}$.

We let $\underline{\underline{G}}_{\mathbf{R}}$ be G. As a Lie group G needs not be connected, but it is assumed that $Ad_G G$ is contained in the connected complex adjoint group of $\mathfrak{g}_C$. Then G satisfies the assumptions in Harish-Chandra [3, pp. 105-106]. For Theorem 1.4.1, which will be needed in the proof of Selberg's principle, we restrict $\underline{\underline{G}}$ further so that its complex Lie algebra is a complex *classical semi-simple Lie algebra* (cf. Varadarajan [1, pp. 292-305]). Then the real Lie algebra of G is a real classical semi-simple Lie algebra. For a list of examples of $\underline{\underline{G}}$ and G, the reader may, for instance, refer to Knapp [1, pp. 4-6].

Let Γ be an *arithmetic subgroup* of G, i.e., Γ is a subgroup of G such that

1) it is contained in G_Q ;

2) both $[\Gamma: \Gamma \cap G_Z]$ and $[G_Z: \Gamma \cap G_Z]$ are finite.

In the case of $G = SL(2, \mathbf{R})$, this definition encompasses the modular group, the discrete subgroups of Hecke's theory of modular forms, and the congruence subgroups. In the present generality this definition is motivated by the fact that these groups form the natural class of discrete subgroups to which modern reduction theory, as developed by Godement, Borel, and Harish-Chandra, applies (Borel [1, p. 51]).

Under these hypotheses, $\mathrm{Vol}(\Gamma \setminus G) < \infty$.

Presently we make one more assumption on G and Γ, namely that there exists a Siegel domain S_0 such that $G = \Gamma S_0$. We will elaborate on this in Section 2.4. The most elementary example is:

$$\underline{\underline{G}} = SL(n, \mathbf{C}), \quad G = SL(n, \mathbf{R}), \quad \Gamma = SL(n, \mathbf{Z}) .$$

We now introduce some notations which will be used throughout this work.

Since $G \subset GL(n, \mathbf{R})$, each element of G has a *norm*, namely, its usual matrix norm:

$$\| g \| = \left(\sum_{i,j} a_{ij}^2 \right)^{1/2} \qquad \text{for } g = (a_{ij}) \in G .$$

The Lie algebra of G will be denoted by $\boldsymbol{g}$.

Fix a *maximal compact subgroup* K of G, whose Lie algebra will be denoted by $\boldsymbol{k}$. Although G may not be connected, it has the property that K meets each of its connected components. Let σ be a *character* of K. A (complex-valued) function f on G is called a *σ-function* if

$$f(gk) = \sigma(k)f(g) \quad \text{for all } g \in G,\ k \in K\,.$$

We next consider cuspidal subgroups of G. Denote by $\underline{\underline{G}}^{o}$ *the connected component* of $\underline{\underline{G}}$. A *Borel subgroup* of $\underline{\underline{G}}^{o}$ is a maximal connected closed solvable subgroup of $\underline{\underline{G}}^{o}$ (Borel [2]). A *parabolic* **Q**-*subgroup* of $\underline{\underline{G}}^{o}$ is a **Q**-subgroup of $\underline{\underline{G}}^{o}$ which contains a Borel subgroup of $\underline{\underline{G}}^{o}$. Let $\underline{\underline{P}}$ be the normalizer in $\underline{\underline{G}}$ of a parabolic **Q**-subgroup of $\underline{\underline{G}}^{o}$, and set $P = \underline{\underline{P}}_{\mathbf{R}}$. Then P is called a *cuspidal subgroup* of G (Harish-Chandra [2, pp. 2-3], Langlands [3, p. 235]; cf. also Langlands [1, p. 15]). Every cuspidal subgroup P has a *Langlands decomposition:* $P = NAM$. If Γ is an arithmetic subgroup of G, then

$$\Gamma \cap P = \Gamma \cap NM\,.$$

Moreover, $\Gamma \cap N \setminus N$ is compact. A cuspidal subgroup of G is a parabolic subgroup of G as a real reductive Lie group.

Fix a *minimal cuspidal subgroup* P_0. We have $G = P_0K$, and P_0 has a Langlands decomposition: $P_0 = N_0A_0M_0$. The corresponding Lie algebra will be denoted by $\boldsymbol{p}_0, \boldsymbol{n}_0, \boldsymbol{a}_0, \boldsymbol{m}_0$. Fix a set of *simple roots* of $(\boldsymbol{p}_0, \boldsymbol{a}_0)$ and denote it by Σ^o. We set

$$\Sigma^o = \{\alpha_r : r = 1, \dots, m\}, \qquad m = \dim \boldsymbol{a}_0\,.$$

For every $F \subset \Sigma^o$ there corresponds a *standard cuspidal subgroup* P_F, $P_F \supset P_0$. We have $G = P_FK$, and P_F has the Langlands decomposition: $P_F = N_FA_FM_F$. The corresponding Lie algebra will be denoted by $\boldsymbol{p}_F$, $\boldsymbol{n}_F$, $\boldsymbol{a}_F$, $\boldsymbol{m}_F$. In fact,

$$\boldsymbol{a}_F = \{H \in \boldsymbol{a}_0 : \alpha(H) = 0 \quad \forall\, \alpha \in F\}\,.$$

If $F \subset F' \subset \Sigma^o$, then

$$P_F \subset P_{F'}, \qquad N_F \supset N_{F'}, \qquad A_F \supset A_{F'}, \qquad M_F \subset M_{F'}\,.$$

For $F_r = \Sigma^o \setminus \{\alpha_r\}$, we call $P_r = P_{F_r}$ the rth *maximal standard cuspidal subgroup.* We have $G = P_rK$, and P_r has a Langlands decomposition: $P_r = N_rA_rM_r$. The corresponding Lie algebras will be denoted by $\boldsymbol{p}_r$, $\boldsymbol{n}_r$, $\boldsymbol{a}_r$, $\boldsymbol{m}_r$. It is clear that dim $\boldsymbol{a}_r = 1$.

Only complex scalar-valued functions will be considered in this work.

For a given arithmetic subgroup Γ, if f is a square integrable or continuous function on $\Gamma \setminus G$, the *constant term* of f with respect to P_r is defined by:

$$f^r(g) = \int_{N_r \cap \Gamma \setminus N_r} f(ng)\, dn \qquad (g \in G) .$$

(The measure on $N_r \cap \Gamma \setminus N_r$ is normalized so that $N_r \cap \Gamma \setminus N_r$ has measure 1.) More generally, one can define the constant term of f with respect to any maximal cuspidal subgroup. A square integrable function on $\Gamma \setminus G$ is called a *cusp form* if its constant term with respect to any maximal cuspidal subgroup is 0 for almost all $g \in G$.

For the definition of an Eisenstein series for a maximal standard cuspidal subgroup, we fix r, say, to be 1. Most of the time we will drop the subscript "1" and write P for P_1. We have $G = PK$; $P = NAM$ with the corresponding Lie algebras $\boldsymbol{p}$, $\boldsymbol{n}$, $\boldsymbol{a}$, $\boldsymbol{m}$.

The *Levi component* M has properties analogous to those of G. For instance, its Lie algebra is a classical semi-simple Lie algebra. One particular arithmetic subgroup of M can be derived from Γ,

$$\Gamma_M = \pi_M(\Gamma \cap NM) ,$$

where $\pi_M \colon P \to M$ is the projection map according to the Langlands decomposition. It can be shown that $\mathrm{Vol}(\Gamma_M \setminus M) < \infty$. One particular maximal compact subgroup of M can be derived from K:

$$K_M = K \cap M .$$

The maximal compact subgroup K_M meets every connected component of M. One particular character of K_M can be derived from σ:

$$\sigma_M(k) = \sigma(k) \quad (k \in K_M)\,.$$

If Φ is a σ_M-function on M, Φ can be extended to a σ-function on G by

$$\Phi(g) = \Phi(namk) = \sigma(k)\Phi(m) \qquad (g = namk,\ \ n \in N,\ \ a \in A,\ \ m \in M,\ \ k \in K)\,.$$

The group G is, among other things, a *real reductive Lie group*. If $\tilde{G}$ is any real reductive Lie group, with a maximal compact subgroup K, that satisfies the assumption in Harish-Chandra [3, pp. 105-106] (hereafter these assumptions will always be in force when dealing with a real reductive Lie group), we denote by $\boldsymbol{D}(\tilde{G})$, $\boldsymbol{D}_K(\tilde{G})$, and $\mathbf{Z}(\tilde{G})$ respectively the *left* $\tilde{G}$*-invariant differential operators* on $\tilde{G}$, the *left* $\tilde{G}$*-invariant and right* K*-invariant differential operators* on $\tilde{G}$, and the *bi-invariant differential operators* on $\tilde{G}$. A function f on $\tilde{G}$ is said to be smooth if $f \in C^\infty(\tilde{G})$. A smooth function f is said to be $\mathbf{Z}(\tilde{G})$*-finite* if $\{Df : D \in \mathbf{Z}(\tilde{G})\}$ is a finite-dimensional vector space. For example, an eigenfunction for all of $\mathbf{Z}(\tilde{G})$ is $\mathbf{Z}(\tilde{G})$-finite. If f is a function on $\tilde{G}$, and $z \in \tilde{G}$, let ${}^z f$ be the function defined by

$${}^z f(g) = f(zg) \qquad (g \in \tilde{G})\,,$$

and f^z be defined by

$$f^z(g) = f(gz) \qquad (g \in \tilde{G})\,.$$

For a smooth function f,

$$D\,{}^z f = {}^z(Df) \quad \text{for all } D \in \boldsymbol{D}(\tilde{G})\,.$$

A function f is said to be K-*finite* if the span of $\{f^k : k \in K\}$ is finite-dimensional. It is clear that a σ-function is K-finite.

1.3. The Eisenstein Series

We can now define the Eisenstein series whose analytic continuation is the subject of this work. The main reference of this section is Harish-Chandra [2], Chapter 2, Sections 1-2.

Let ${}^{o}L^2(\Gamma_M \backslash M, \sigma_M, \chi)$ denote the vector space of all cusp forms on M which are σ_M-functions and are eigenfunctions for all of $\mathbf{Z}(M)$, with eigenvalues $\chi(D_M)$ $(D_M \in \mathbf{Z}(M))$.

For the definition of an Eisenstein series fix a nonzero $\Phi \in {}^{o}L^2(\Gamma_M \backslash M, \sigma_M, \chi)$. Let Φ be extended to a σ-function on G, which we still denote by Φ; and set

$$\Phi_\Lambda(g) = \Phi_\Lambda(namk) = \Phi(g)e^{(-\Lambda-\rho)(\log a)}$$

$$= \sigma(k)\Phi(m)e^{(-\Lambda-\rho)(\log a)} \quad ((\Lambda, g) \in \boldsymbol{a}_C^* \times G),$$

where $\boldsymbol{a}_C^*$ denotes the dual of $\boldsymbol{a}_C$, the complexified Lie algebra; ρ is one-half the sum of the positive roots of $(\boldsymbol{p}, \boldsymbol{a})$. Now define the *cuspidal Eisenstein series* $E(\Lambda, \Phi, g)$ for the maximal standard cuspidal subgroup P to be

$$E(\Lambda, \Phi, g) = \sum_{\gamma \in \Gamma \cap P \backslash \Gamma} \Phi_\Lambda(\gamma g).$$

Let α be the simple root of $(\boldsymbol{p}, \boldsymbol{a})$ and $H \in \boldsymbol{a}$ be such that $\alpha(H) > 0$. Set

$$(\boldsymbol{a}_C^*)^+ = \{\Lambda \in \boldsymbol{a}_C^*: \operatorname{Re}(\Lambda)(H) - \rho(H) > 0\}.$$

For every compact $V \subset (\boldsymbol{a}_C^*)^+$ and every compact $C \subset G$, the Eisenstein series converges absolutely and uniformly on $V \times C$.

The following is a list of the basic properties of the Eisenstein series:

1) For each $\Lambda \in (\boldsymbol{a}_C^*)^+$, it defines a Γ-*automorphic* function on G, i.e., $E(\Lambda, \Phi, \gamma g) = E(\Lambda, \Phi, g)$ for all $\gamma \in \Gamma$.

2) It is a C^∞-function on $(\boldsymbol{a}_C^*)^+ \times G$, and is holomorphic on $(\boldsymbol{a}_C^*)^+$ for each $g \in G$.

3) The series can be differentiated term by term with respect to (Λ, g).

4) The Eisenstein series is an eigenfunction for all of $\mathbf{Z}(G)$:

$$DE = \chi_\Lambda(D)E \quad (D \in \mathbf{Z}(G)),$$

where

$$\chi_\Lambda(D) = \chi(\mu_\Lambda(D)), \quad \mu_\Lambda\colon \mathbf{Z}(G) \to \mathbf{Z}(M),$$

and $\chi_\Lambda(D)$ is an entire function for $\boldsymbol{a}_C^*$ for each $D \in \mathbf{Z}(G)$.

5) For each $\Lambda \in (\boldsymbol{a}_C^*)^+$, it is a σ-function on G.

We end this section with a remark on notation. The Eisenstein series has been constructed with a cusp form Φ on M, and is dependent on the group variable g and $\Lambda \in (\boldsymbol{a}_C^*)^+$. Depending on which parameters we wish to emphasis, we will write:

$$E(\Lambda, \Phi, g), \quad E(\Lambda, g), \quad E(\Lambda), \quad E(g), \quad E\,.$$

1.4. Selberg's Eigenfunction Principle

Let α be a compactly supported continuous function on G. For $f \in C(G)$, the *convolution* $\alpha * f$ is define by:

$$(\alpha * f)(g) = \int_G \alpha(h^{-1}g)f(h)\,dh \qquad (g \in G)\,.$$

If Ω denotes the compact support of α, then

$$(\alpha * f)(g) = \int_{g\Omega^{-1}} \alpha(h^{-1}g)f(h)\,dh \qquad (g \in G)\,.$$

Note that each $g\Omega^{-1}$ is compact.

As before,

$$\Phi_\Lambda(g) = \Phi(g)e^{(\Lambda-\rho)(\log a)} = \sigma(k)\Phi(m)e^{(\Lambda-\rho)(\log a)}$$

($g = namk$, $n \in N$, $a \in A$, $m \in M$, $k \in K$), where $\Phi \in {}^0L^2(\Gamma_M \backslash M, \sigma_M, \chi)$ has been extended to a σ-function on G, also denoted by Φ, according to $\Phi(namk) = \sigma(k)\Phi(m)$. The Eisenstein series

$$E(\Lambda, \Phi, g) = \sum_{\Gamma \cap P \backslash \Gamma} \Phi_\Lambda(\gamma g)$$

converges absolutely and uniformly on $V \times C$ for any compact subsets V and C of $(\mathfrak{a}_C^*)^+$ and G respectively.

If α is any *continuous compactly supported* function on G such that

$$\alpha(k^{-1}gk) = \alpha(g) \quad \text{for all } g \in G, k \in K,$$

then

$$\alpha * \Phi_\Lambda = \hat{\alpha}^o(\Lambda)\Phi_\Lambda,$$

where, for a given σ, χ, and Λ, $\hat{\alpha}^o(\Lambda)$ is a constant function on G independent of Φ. This complete assertion is our goal. It will be placed in the context of Selberg's principle.

Suppose it has been proved, then since

$$\alpha * E = \alpha * \sum_{\Gamma \cap P \backslash \Gamma} {}^\gamma\Phi_\Lambda$$

$$= \sum_{\Gamma \cap P \backslash \Gamma} \alpha * {}^\gamma\Phi_\Lambda$$

and

$$(\alpha * {}^\gamma\Phi_\Lambda)(g) = \int_G \alpha(h^{-1}g)\Phi_\Lambda(\gamma h)\, dh,$$

which becomes

$$\int_G \alpha(h^{-1}\gamma g)\Phi_\Lambda(h)\, dh = (\alpha * \Phi_\Lambda)(\gamma g)$$

under the change of variables $\gamma h \mapsto h$, we have

$$\alpha * E = \hat{\alpha}^o(\Lambda)E.$$

The convolution $\alpha * \Phi_\Lambda$ can be reduced to a convolution of Φ on the level of the Levi component M. Recall that

$$(\alpha * \Phi_\Lambda)(g) = \int_G \alpha(g'^{-1}g)\Phi_\Lambda(g')\, dg' .$$

For $g = namk$, $g' = n'a'm'k'$,

$$(\alpha * \Phi_\Lambda)(g) = \int_M \int_A \int_N \int_K \alpha(k'^{-1}m'^{-1}a'^{-1}n'^{-1}namk)\, \sigma(k')\Phi(m')e^{(-\Lambda-\rho)(\log a')}e^{2\rho(\log a')}\, dk'dn'da'dm'$$

$$= \int_M \left[\int_A \int_N \int_K \alpha(k'^{-1}m'^{-1}a'^{-1}n'^{-1}namk)\, \sigma(k')e^{(-\Lambda-\rho)(\log a')}e^{2\rho(\log a')}\, dk'dn'da' \right] \Phi(m')\, dm' .$$

Now

$$\alpha(k'^{-1}m'^{-1}a'^{-1}n'^{-1}namk) = \alpha(m'^{-1}a'^{-1}n'^{-1}namkk'^{-1})$$

$$= \alpha(m'^{-1}a'^{-1}n'^{-1}n(a'a'^{-1})a(m'm'^{-1})mkk'^{-1})$$

$$= \alpha(m'^{-1}a'^{-1}n'^{-1}n(a'm'a'^{-1}am'^{-1})mkk'^{-1})$$

$$= \alpha((a'm')^{-1}(n^{-1}n')^{-1}(a'm')(a^{-1}a')^{-1}m'^{-1}m(k'k^{-1})^{-1}) .$$

We shall make several changes of variable to transform the triple inner integral:

$$\int_K \alpha((a'm')^{-1}(n^{-1}n')^{-1}(a'm')(a^{-1}a')^{-1}m'^{-1}m(k'k^{-1})^{-1})\, \sigma(k')e^{(-\Lambda-\rho)(\log a')}e^{2\rho(\log a')}\, dk' ,$$

which becomes, under $k'k^{-1} \mapsto k'$,

$$\int_K \alpha((a'm')^{-1}(n^{-1}n')^{-1}(a'm')(a^{-1}a')^{-1}m'^{-1}mk'^{-1})\, \sigma(k'k)e^{(-\Lambda-\rho)(\log a')}e^{2\rho(\log a')}\, dk'$$

$$= \sigma(k)\int_K \alpha((a'm')^{-1}(n^{-1}n')^{-1}(a'm')(a^{-1}a')^{-1}m'^{-1}mk'^{-1})\, \sigma(k')e^{(-\Lambda+\rho)(\log a')}\, dk' .$$

Next,

$$\sigma(k)\int_N \alpha((a'm')^{-1}(n^{-1}n')^{-1}(a'm')(a^{-1}a')^{-1}m'^{-1}mk'^{-1})\, \sigma(k')e^{(-\Lambda+\rho)(\log a')}\, dn' ,$$

under $n^{-1}n' \mapsto n'$ and $n'^{-1} \mapsto n'$,

$$= \sigma(k)\int_N \alpha((a'm')^{-1}n'(a'm')(a^{-1}a')^{-1}m'^{-1}mk'^{-1})\, \sigma(k')e^{(-\Lambda+\rho)(\log a')}\, dn' ,$$

under $(a'm')^{-1}n'(a'm') \mapsto n'$,

$$= \sigma(k)\int_N \alpha(n'(a^{-1}a')^{-1}m'^{-1}mk'^{-1})\,\sigma(k')e^{(-\Lambda+\rho)(\log a')}e^{-2\rho(\log a')}\,dn'$$

$$= \sigma(k)\int_N \alpha(n'(a^{-1}a')^{-1}m'^{-1}mk'^{-1})\sigma(k')e^{(-\Lambda-\rho)(\log a')}\,dn'\,.$$

Finally,

$$\sigma(k)\int_A \alpha(n'(a^{-1}a')^{-1}m'^{-1}mk'^{-1})\sigma(k')e^{(-\Lambda-\rho)(\log a')}\,da'\,,$$

under $a^{-1}a' \mapsto a'$,

$$= \sigma(k)\int_A \alpha(n'a'^{-1}m'^{-1}mk'^{-1})\sigma(k')e^{(-\Lambda-\rho)(\log aa')}\,da'$$

$$= \sigma(k)\left[\int_A \alpha(n'a'^{-1}m'^{-1}mk'^{-1})\sigma(k')e^{(-\Lambda-\rho)(\log a')}\,da'\right]e^{(-\Lambda-\rho)(\log a)}\,,$$

under $a'^{-1} \mapsto a'$,

$$= \sigma(k)\left[\int_A \alpha(n'a'm'^{-1}mk'^{-1})\sigma(k')e^{(\Lambda+\rho)(\log a')}\,da'\right]e^{(-\Lambda-\rho)(\log a)}\,.$$

Therefore the convolution

$$(\alpha * \Phi_\Lambda)(g) = \sigma(k)(\alpha_\Lambda * \Phi)(m)e^{(-\Lambda-\rho)(\log a)}\,, \tag{1.4.1}$$

where

$$\alpha_\Lambda(m) = \int_A\int_N\int_K \alpha(n'a'mk'^{-1})\sigma(k')e^{(\Lambda+\rho)(\log a')}\,dk'dn'da' \qquad (m \in M)\,.$$

Lemma 1.4.1. The function α_Λ is continuous and compactly supported on M.

Proof. The continuity of α_Λ can be verified in a straightforward manner. Denote by Ω the compact support of α. For $n' \in N$, $a' \in A$, $m \in M$, $k'^{-1} \in K$, $\alpha(n'a'mk'^{-1}) \neq 0$ implies $n'a'mk'^{-1} \in \Omega$ implies $n'a'm \in \Omega K \cap P$. Both P and ΩK are closed. The set $\Omega K \cap P$ is a closed subset of ΩK, which is compact. Hence $\Omega K \cap P$ is a compact subset of P. If $\pi_M: P \to M$ denotes the continuous projection from P into M, then $m \in \pi_M(\Omega K \cap P)$, which is a compact subset of M. Therefore, α_Λ is compactly supported. □

Although α_Λ^* actually maps $L^2(\Gamma_M \setminus M, \sigma_M, \chi)$ into itself, Proposition 1.4.1 suffices for our development. Denote by $C^\infty(M, \sigma_M, \chi)$ the set of all smooth functions on M which are σ_M-functions and are eigenfunctions for all of $\mathbf{Z}(M)$, with eigenvalues $\chi(D_M)$ $(D_M \in \mathbf{Z}(M))$.

Proposition 1.4.1. The mapping α_Λ^* sends $C^\infty(M, \sigma_M, \chi)$ into itself.

Proof. For $k \in K_M$,

$$(\alpha_\Lambda * \Phi)(mk) = \int_M \alpha_\Lambda(m'^{-1}mk)\Phi(m')\, dm'$$

$$= \int_M \int_A \int_N \int_K \alpha(n'a'm'^{-1}mkk'^{-1})\sigma(k')e^{(\Lambda+\rho)(\log a')}\Phi(m')\, dk'dn'da'dm'$$

$$= \int_M \int_A \int_N \int_K \alpha(n'a'm'^{-1}m(k'k^{-1})^{-1})\sigma(k')e^{(\Lambda+\rho)(\log a')}\Phi(m')\, dk'dn'da'dm' ,$$

under $k'k^{-1} \mapsto k'$,

$$= \int_M \int_A \int_N \int_K \alpha(n'a'm'^{-1}mk'^{-1})\sigma(k'k)e^{(\Lambda+\rho)(\log a')}\Phi(m')\, dk'dn'da'dm'$$

$$= \sigma_M(k)\int_M \int_A \int_N \int_K \alpha(n'a'm'^{-1}mk'^{-1})\sigma(k')e^{(\Lambda+\rho)(\log a')}\Phi(m')\, dk'dn'da'dm'$$

$$= \sigma_M(k)(\alpha_\Lambda * \Phi)(m) .$$

So $\alpha * \Phi$ is a σ_M-function.

On the other hand,

$$(\alpha_\Lambda * \Phi)(m) = \int_M \alpha_\Lambda((m^{-1}m')^{-1})\Phi(m')\, dm' ,$$

under $m^{-1}m' \mapsto m'$ and $m'^{-1} \mapsto m'$,

$$= \int_M \alpha_\Lambda(m')\Phi(mm'^{-1})\, dm' .$$

Hence,

$$[D(\alpha_\Lambda * \Phi)](m) = \int_M \alpha_\Lambda(m')(D\Phi)(mm'^{-1})\, dm'$$

$$= \chi(D)\int_M \alpha_\Lambda(m')\Phi(mm'^{-1})\, dm' = \chi(D)(\alpha_\Lambda * \Phi)(m) \quad \text{for } D \in \mathbf{Z}(M) .$$

This completes the proof that $\alpha_\Lambda * \Phi$ belongs to $C^\infty(M, \sigma_M, \chi)$ as Φ does. □

Recall that Φ is a σ_M-function on M, and is an eigenfunction for all the bi-invariant differential operators on M. In view of the equation (1.4.1) our goal will be reached by establishing the following fundamental principle:

Let $\tilde{G}$ be a real reductive Lie group with a maximal compact subgroup K, σ a character of K; *if* Φ *is a* σ*-function on* $\tilde{G}$ *and is an eigenfunction for all the bi-invariant differential operators on* $\tilde{G}$*, so that*

$$D\Phi = \chi(D)\Phi \quad \textit{for all } D \in \mathbf{Z}(\tilde{G}) ,$$

α is a continuous compactly supported function on $\tilde{G}$ *such that* $\alpha * \Phi$ *is also a* σ*-function,* then there exists a constant function $\hat{\alpha}$ on $\tilde{G}$, which, for a given σ and a given χ, does not depend on Φ, such that

$$\alpha * \Phi = \hat{\alpha}\Phi .$$

This falls within the framework of the so-called *Selberg's principle*, and $\hat{\alpha}$ is a *Selberg transform* of α. We shall establish this particular case under the additional assumption that the Lie algebra of $\tilde{G}$ is a real classical semi-simple Lie algebra.

Let $\boldsymbol{D}(\tilde{G})$, $\boldsymbol{D}_K(\tilde{G})$ and $\mathbf{Z}(\tilde{G})$ be respectively the left $\tilde{G}$-invariant differential operators on $\tilde{G}$, the left $\tilde{G}$-invariant and right K-invariant differential operators on $\tilde{G}$, and the bi-invariant differential operators on $\tilde{G}$. It is clear that

$$\mathbf{Z}(\tilde{G}) \subset \boldsymbol{D}_K(\tilde{G}) \subset \boldsymbol{D}(\tilde{G}) .$$

The motivation for the fundamental principle is that any smooth function on $\tilde{G}$ which is $\mathbf{Z}(\tilde{G})$-finite and right K-finite is real-analytic (Baily [1, p. 154]) and that, by virtue of *Taylor's formula* for analytic functions on a Lie group (Helgason [1, p. 95]), two analytic σ-functions f_1 and f_2 on $\tilde{G}$ are equal if at some point $g \in \tilde{G}$,

$$(Df_1)(g) = (Df_2)(g) \quad \text{for all } D \in \boldsymbol{D}(\tilde{G}) .$$

The second assertion does follow because the fact that K meets every connected component of $\tilde{G}$ implies that K acts transitively on the set of components of $\tilde{G}$, hence a σ-function is completely determined by its values on any single component. Note that in the present situation both Φ and $\alpha * \Phi$ are analytic.

Our first step is to show that Φ is actually an eigenfunction for all of $\boldsymbol{D}_K(\tilde{G})$. For this purpose we present a theorem on the structure of $\boldsymbol{D}_K(\tilde{G})$. Let $\tilde{\boldsymbol{g}}$ and $\boldsymbol{k}$ be respectively the Lie algebra of $\tilde{G}$ and K; $U(\tilde{\boldsymbol{g}})$, $U(\tilde{\boldsymbol{g}})^K$, and $Z(\tilde{\boldsymbol{g}})$ be respectively the universal enveloping algebra of $\tilde{\boldsymbol{g}}_C$, the set of all $Ad_{\tilde{g}}(K)$-invariant elements of $U(\tilde{\boldsymbol{g}})$, and the center of $U(\tilde{\boldsymbol{g}})$. Then $\boldsymbol{D}(\tilde{G})$, $\boldsymbol{D}_K(\tilde{G})$, and $\mathbf{Z}(\tilde{G})$ can respectively be identified with $U(\tilde{\boldsymbol{g}})$, $U(\tilde{\boldsymbol{g}})^K$, and $Z(\tilde{\boldsymbol{g}})$ (Helgason [1, pp. 385-393]).

<u>Theorem 1.4.1.</u> Let $\tilde{G}$ be a real reductive Lie group with a maximal compact subgroup K. Suppose $\tilde{G}$, the Lie algebra of $\tilde{G}$, is a real classical semi-simple Lie algebra, then

$$U(\boldsymbol{g})^K = Z(\tilde{\boldsymbol{g}}) + U(\tilde{\boldsymbol{g}})^K \cap U(\tilde{\boldsymbol{g}})\boldsymbol{k}_C .$$

Moreover, if

$$u = u_Z + u^K \quad \text{for } u \in U(\tilde{\boldsymbol{g}})^K ,$$

where $u_Z \in Z(\tilde{\boldsymbol{g}})$, $u^K \in U(\tilde{\boldsymbol{g}})^K \cap U(\tilde{\boldsymbol{g}})\boldsymbol{k}_C$, u_Z and u^K can be chosen so that

$$\deg(u_Z), \deg(u^K) \le \deg(u) .$$

Some background is needed for the proof of Theorem 1.4.1. We shall recall certain descriptions of the subalgebras $U(\tilde{g})^K$ and $Z(\tilde{g})$ of $U(\tilde{g})$ respectively in terms of direct sum decompositions of $\tilde{g}$ and $\tilde{g}_C$. Both of these descriptions are well known for real reductive Lie groups. We will also use a theorem of Helgason about classical semi-simple Lie algebras.

Let $\tilde{G} = NAK$ be an *Iwasawa decomposition* of $\tilde{G}$, with the corresponding Lie algebras of $\tilde{g}$, $\boldsymbol{n}$, $\boldsymbol{a}$, $\boldsymbol{k}$. We have

$$\tilde{g} = \boldsymbol{n} \oplus \boldsymbol{a} \oplus \boldsymbol{k}\,.$$

From the *Poincaré-Birkhoff-Witt theorem,*

$$U(\tilde{g}) = \boldsymbol{k}_C U(\tilde{g}) \oplus U(\boldsymbol{a}) \oplus U(\tilde{g})\boldsymbol{n}_C\,,$$

where $U(\boldsymbol{a})$ denotes the universal enveloping algebra of $\boldsymbol{a}_C$. Let δ_K be the restriction to $U(\tilde{g})^K$ of the projection map from $U(\tilde{g})$ onto $U(\boldsymbol{a})$ with respect to this decomposition. Extend $\boldsymbol{a}$ to a Cartan subalgebra $\boldsymbol{h}$ of $\tilde{g}$ such that $\boldsymbol{h} = \boldsymbol{h}_I \oplus \boldsymbol{a}$, where $\boldsymbol{h}_I$ and $\boldsymbol{a}$ are orthogonal with respect to the Killing form on $\tilde{g} \times \tilde{g}$. Denote by ρ one-half of the sum of positive roots of $(\tilde{g}, \boldsymbol{a})$, and let $\eta_K(v) = v - \rho(v)$ for $v \in \boldsymbol{a}$. The map η_K extends to an algebra automorphism of $U(\boldsymbol{a})$, which will also be denoted by η_K. Set $\gamma_K = \eta_K \circ \delta_K$. Denote the Weyl group of $(\tilde{g}, \boldsymbol{a})$ by W_K. The following properties of γ_K are elementary (cf. Helgason [2, pp. 305-305]; cf. also Harish-Chandra [4], Schlichtkrull [1, pp. 60-61]): it is an algebra homomorphism from $U(\tilde{g})^K$ into $U(\boldsymbol{a})^{W_K}$, *the* W_K-*invariant elements of* $U(\boldsymbol{a})$; It preserves degree, i.e., $\deg(\gamma_K(u)) \le \deg(u)$; and it has kernel

$$U(\tilde{g})^K \cap U(\tilde{g})\boldsymbol{k}_C\,.$$

On the other hand, denoting the Weyl group of $(\tilde{g}_C, \boldsymbol{h}_C)$ by W, there exists an algebra isomorphism γ from $Z(\tilde{g})$ onto $U(\boldsymbol{h})^W$. This map is defined as follows (Harish-Chandra [5, 6]; also Knapp [1, pp. 218-223]). Let P denote the set of positive roots of $(\tilde{g}_C, \boldsymbol{h}_C)$, and set

$$\boldsymbol{n}_C^+ = \sum_{\alpha \in P} \tilde{g}_C^{\alpha}\,, \quad \boldsymbol{n}_C^- = \sum_{\alpha \in P} \tilde{g}_C^{-\alpha}\,.$$

Then

$$\tilde{g}_C = n_C^+ \oplus h_C \oplus n_C^- ,$$

and one can deduce form this that

$$Z(\tilde{g}) \subset U(h) \oplus U(\tilde{g})n_C^+ .$$

Let δ be the projection map from $Z(\tilde{g})$ into $U(h)$. Denote by ρ_C one-half of the sum of positive roots of $(\tilde{g}_C, h_C)$, and let $\eta(H) = H - \rho_C(H)$ for $H \in h_C$. The map η extends to an automorphism of $U(h)$. We define γ to be $\eta \circ \delta$.

Since a (respectively h) is abelian, the *symmetrization mapping* identifies $U(a)$ ($U(h)$) with the *symmetric algebra* $S(a_C)$ ($S(h_C)$). Under this identification $U(a)^{W_K}$ ($U(h)^W$) corresponds to a set of W_K-invariant (W-invariant) polynomials $I(a_C)$ ($I(h_C)$). We may regard γ_K as mapping $U(\tilde{g})^K$ into $I(a_C)$ (respectively γ as mapping $Z(\tilde{g})$ onto $I(h_C)$).

For $z \in Z(\tilde{g})$, $\gamma_K(z)$ and $\gamma(z)$ are related as follows. Regard a_C^* as contained in h_C^* by setting every $\Lambda \in a_C^*$ equal to 0 on h_I. We may define $\rho_I = \rho_C - \rho$. Then one can show that

$$\gamma(z)(\Lambda - \rho_I) = \gamma_K(z)(\Lambda) \quad \text{for all } \Lambda \in a_C^* .$$

We reinterpret $I(a_C)$ and $I(h_C)$. By virtue of the restriction of the Killing form on $\tilde{g}_C \times \tilde{g}_C$ to $a_C \times a_C$ (respectively $h_C \times h_C$), the elements of $I(a_C)$ ($I(h_C)$) can be regarded as polynomial functions on a_C (h_C). One then deduces easily from the last relation that

$$\deg(\gamma(z)^o - \gamma_K(z)) < \deg(z) ,$$

where $\gamma(z)^o$ is the restriction of $\gamma(z) \in I(h_C)$ to a_C.

By restricting the elements of $I(h_C)$, which are polynomial functions on h_C, to a_C, we obtain a map from $I(h_C)$ into $I(a_C)$. Helgason [3] proved that if $\tilde{g}$ is a classical semi-simple Lie group, then this map is surjective. He accomplished this through a case-by-case verification. In the same reference, counterexamples were also given if the restriction on $\tilde{g}$ is removed.

We now present the proof of the theorem.

Proof of Theorem 1.4.1. We proceed by induction on $\deg(\gamma_K(u))$ for $u \in U(\tilde{g})^K$ to prove that for every $u \in U(\tilde{g})^K$ there exist $u_Z \in Z(\tilde{g})$ and $u^K \in U(\tilde{g})^k \cap U(\tilde{g})\boldsymbol{h}_C$ such that

$$\deg(u_Z),\ \deg(u^K) \le \deg(u)$$

and

$$u = u_Z + u^K .$$

Consider first the case $\deg(\gamma_K(u)) = 0$. There exists $c \in \mathbf{C}$ such that $\gamma_K(u)$ is of constant value c on $\boldsymbol{a}_C$. By abuse of notation, c will also denote an element in $Z(\tilde{g})$ and an element in $I(\boldsymbol{a}_C)$. It is easily checked that $\gamma_K(c) = c$. Hence $\gamma_K(u - c) = 0$, i.e., $u - c$ is in the kernel of γ_K, which is $U(\tilde{g})^K \cap U(\tilde{g})\boldsymbol{k}_C$. The desired result then clearly follows.

Now suppose the assertion to be proved is true for $0 \le \deg(\gamma_K(u)) \le n$. Let $u \in U(\tilde{g})^K$ be such that $\deg(\gamma_K(u)) = n + 1$. By Helgason's theorem, there exists $p \in I(\boldsymbol{h}_C)$ whose restriction to $\boldsymbol{a}_C$ is equal to $\gamma_K(u)$. Since every polynomial function in $I(\boldsymbol{h}_C)$ is a sum of homogeneous elements of $I(\boldsymbol{h}_C)$ of distinct degrees, it may be assumed that p has the same degree as $\gamma_K(u)$. Now there exists $z \in Z(\tilde{g})$ such that $\deg(z) = \deg(p) = n + 1$ and $\gamma(z) = p$. Hence $\gamma(z)^o = \gamma_K(u)$. Then $\deg(\gamma(z)^o - \gamma_K(z)) < \deg(z)$ implies $\deg(\gamma_K(u - z)) \le n$. Of course, $u - z \in U(\tilde{g})^K$. So the induction hypothesis implies the existence of $z' \in Z(\tilde{g})$ and $u^K \in U(\tilde{g})^K \cap U(\tilde{g})\boldsymbol{k}_C$ such that

$$\deg(z'), \deg(u^{K)} \le \deg(u - z)$$

and

$$u - z = z' + u^K .$$

Since $\deg(z) = \deg(\gamma_K(u)) \le \deg(u)$, we have $\deg(u - z) \le \deg(u)$. Let $u_Z = z + z'$, and the desired result follows. □

By Theorem 1.4.1, if $u \in U(\tilde{g})^K$, then

$$u = u_Z + \sum_{i=1}^{m} u_i X_i ,$$

where $u_Z \in Z(\tilde{g})$; $u_i \in U(\tilde{g})$, $X_i \in \boldsymbol{k}_C$, $m \in \mathbf{Z}^+$, $\sum_{i=1}^{m} u_i X_i \in U(\tilde{g})^K$, and

$$\deg\Big(\sum_{i=1}^{m} u_i X_i \Big) \le \deg(u) .$$

Let $\{Y_1, \ldots, Y_{\beta-1}, Y_\beta, \ldots, Y_n\}$ be a basis of $\tilde{g}$ such that $\{Y_\beta, \ldots, Y_n\}$ is a basis of $\boldsymbol{k}$. Each X_i is a linear combination of $Y_\beta, \ldots, Y_n$. By virtue of Lie bracket relations,

$$(Y_1^{e_1} \cdots Y_p^{e_p} \cdots Y_n^{e_n})Y_p = Y_1^{e_1} \cdots Y_p^{e_p+1} \cdots Y_n^{e_n} + \text{(lower–degree terms)} .$$

Since $\boldsymbol{k}$ is a Lie subalgebra, for $\beta \le p \le n$, $(Y_1^{e_1} \cdots Y_p^{e_p} \cdots Y_n^{e_n})Y_p$ is a linear combination of $Y_1^{e_1} \cdots Y_\beta^{e_\beta} \cdots Y_n^{e_n}$ $(e_\beta + \cdots + e_n > 0)$. Therefore, by regrouping terms in $\sum_{i=1}^{m} u_i X_i$ and dropping extraneous higher-degree terms, we can assume without loss of generality that

$$\deg(u_i) \le \deg(u) - 1 \quad \text{for} \quad i = 1, \ldots, m .$$

It is in this sharpened version that the theorem will be applied. If now $D \in \boldsymbol{D}_K(\tilde{G})$, we can write

$$D = D_Z + \sum_{i=1}^{m} D_i X_i ,$$

where $D_Z \in \mathbf{Z}(\tilde{G})$; $D_i \in \boldsymbol{D}(\tilde{G})$, $X_i \in \boldsymbol{k}_C$, $m \in \mathbf{Z}^+$, $\sum_{i=1}^{m} D_i X_i \in \boldsymbol{D}_K(\tilde{G})$, and

$$\deg(D_i) \le \deg(D) - 1 \quad \text{for} \quad i = 1, \ldots, m .$$

The following corollary to Theorem 1.4.1 shows that in dealing with Φ we can assume without loss of generality that $D_i \in \boldsymbol{D}_K(\tilde{G})$.

Corollary. For $D \in \boldsymbol{D}_K(\tilde{G})$, there exist $D_Z \in \mathbf{Z}(\tilde{G})$, $D_i \in \boldsymbol{D}_K(\tilde{G})$, $X_i \in \boldsymbol{k}_C$, $m \in \mathbf{Z}^+$, such that, for any character σ of K, and any smooth σ-function Φ on $\tilde{G}$,

$$D\Phi = D_Z\Phi + \left[\sum_{i=1}^{m} D_iX_i\right]\Phi$$

and

$$\deg(D_i) \le \deg(D) = 1 \quad \text{for} \quad i = 1, \dots, m.$$

Proof. Let $D_i \in \boldsymbol{D}(\tilde{G})$, $X_i \in \boldsymbol{k}_C$, $m \in \mathbf{Z}^+$ such that $\sum_{i=1}^{m} D_iX_i \in \boldsymbol{D}_K(\tilde{G})$. Since $\sum_{i=1}^{m} D_iX_i \in \boldsymbol{D}_K(\tilde{G})$,

$$\left[\sum_{i=1}^{m} D_iX_i\right]\Phi\,(g) = \int_K \left[\left[\sum_{i=1}^{m} D_iX_i\right]\Phi^k\right](gk^{-1})\,dk$$

$$= \sum_{i=1}^{m} \int_K [(D_iX_i)\Phi^k](gk^{-1})\,dk.$$

Let us consider one summand, and for simplicity drop the subscript "i". We have $X = X_1 + \sqrt{-1}\,X_2$, where $X_1, X_2 \in \boldsymbol{k}$, so

$$\int_K [(DX)\Phi^k](gk^{-1})\,dk = \int_K [D(X_1\Phi^k + \sqrt{-1}\,X_2\Phi^k)](gk^{-1})\,dk.$$

Now

$$(X_1\Phi^k)(z) = \left[\frac{d}{dt}\Phi(z \exp tX_1 \cdot k)\right]_{t=0}$$

$$= \left[\frac{d}{dt}\sigma(k)\sigma(\exp tX_1)\Phi(z)\right]_{t=0}$$

$$= \sigma(k)(X_1\sigma)(e)\Phi(z)$$

$$= (X_1\sigma)(e)\sigma(k)\Phi(z) = (X_1\sigma)(e)\Phi^k(z) .$$

In particular,

$$X_1\Phi = (X_1\sigma)(e)\Phi .$$

Hence,

$$X\Phi = (X\sigma)(e)\Phi .$$

Therefore,

$$\int_K [(DX)\Phi^k](gk^{-1})\, dk = (X\sigma)(e)\int_K (D\Phi^k)(gk^{-1})\, dk$$

$$= (X\sigma)(e)\left[\int_K \mathrm{Ad}(k)D\, dk\right]\Phi\,(g)$$

$$= \left[\int_K \mathrm{Ad}(k)D\, dk\cdot X\right]\Phi\,(g) .$$

Thus,

$$\left[\sum_{i=1}^{m} D_iX_i\right]\Phi = \sum_{i=1}^{m}\left[\int_K \mathrm{Ad}(k)D_i\, dk\cdot X_i\right]\Phi ,$$

where each

$$\int_K \mathrm{Ad}(k)D_i\, dk$$

belongs to $\boldsymbol{D}_K(\tilde{G})$ and has degree less than or equal to that of D_i. □

We now proceed by induction on deg(D) for $D \in \boldsymbol{D}_K(\tilde{G})$ to show that Φ is actually an eigenfunction for all of $\boldsymbol{D}_K(\tilde{G})$.

Theorem 1.4.2. If Φ is a smooth σ-function on $\tilde{G}$, and is an eigenfunction for all of $\mathbf{Z}(\tilde{G})$, so that

$$D\Phi = \chi(D)\Phi \quad \text{for all } D \in \mathbf{Z}(\tilde{G}),$$

then it is an eigenfunction for all of $\boldsymbol{D}_K(\tilde{G})$. Furthermore, for a given σ and a given χ, the eigenvalues for $\boldsymbol{D}_K(\tilde{G})$ are independent of Φ.

Proof. Let $D \in \boldsymbol{D}_K(\tilde{G})$ with $\deg(D) = 1$. By Corollary of Theorem 1.4.1,

$$D\Phi = D_Z\Phi + \left[\sum_{i=1}^{m} C_i X_i\right]\Phi,$$

where $D_Z \in \mathbf{Z}(\tilde{G})$; $C_i \in \mathbf{C}$, $X_i \in k_C$. As

$$X_i\Phi = (X_i\sigma)(e)\Phi,$$

we have

$$D\Phi = \left[\chi(D_Z) + \sum_{i=1}^{m} C_i(X_i\sigma)(e)\right]\Phi.$$

This eigenvalue for D depends on Φ only to the extent of σ and χ.

Now suppose the assertions are true for $1 \le \deg(D) \le n$. Let $D \in \boldsymbol{D}_K(\tilde{G})$ with $\deg(D) = n + 1$. Then by the Corollary

$$D\Phi = D_Z\Phi + \left[\sum_{i=1}^{m} D_i X_i\right]\Phi$$

where $D_i \in \boldsymbol{D}_K(\tilde{G})$ and

$$\deg(D_i) \le \deg(D) - 1 = n.$$

By the induction hypothesis Φ is an eigenfunction for each D_i. Clearly then it is an eigenfunction for D. This eigenvalue for D depends on Φ only to the extent of σ and χ. □

This extension of χ from $\mathbf{Z}(\tilde{G})$ to $\boldsymbol{D}_K(\tilde{G})$ will also be denoted by χ.

Next, following Selberg, the introduction of the *symmetrization operator* will enable us to pass from $\boldsymbol{D}_K(\tilde{G})$ to the whole $\boldsymbol{D}(\tilde{G})$. Given a character σ of K, for every $z \in \tilde{G}$, define an operator

$$M_z : C^\infty(\tilde{G}) \mapsto C^\infty(\tilde{G}) : \Phi \mapsto M_z\Phi$$

by

$$(M_z\Phi)(g) = \int_K \sigma(k^{-1})\Phi(zkz^{-1}g)\, dk \qquad (g \in \tilde{G}) .$$

(The measure on K is normalized so that K has measure 1.) Three properties of M_z are immediate:

1) $D(M_z\Phi) = M_z D\Phi$ for all $D \in \boldsymbol{D}(\tilde{G})$.
2) $(M_z\Phi)(z) = \Phi(z)$ if Φ is a smooth σ-function.
3) $M_z\Phi$ is a σ-function if Φ is a smooth σ-function.

The next lemma shows that M_z commutes with convolution.

Lemma 1.4.2. For any continuous compactly supported function α,

$$M_z(\alpha * \Phi) = \alpha * M_z\Phi .$$

Proof. Starting with the definition of M_z,

$$[M_z(\alpha * \Phi)](g) = \int_K \sigma(k^{-1})(\alpha * \Phi)(zkz^{-1}g)\, dk$$

$$= \int_K \sigma(k^{-1}) \int_{\tilde{G}} \alpha(g'^{-1}zkz^{-1}g)\Phi(g')\, dg'dk$$

$$= \int_K \sigma(k^{-1}) \int_{\tilde{G}} \alpha(g')\Phi(zkz^{-1}gg'^{-1})\, dg'dk ,$$

since $\tilde{G}$ is unimodular. Upon interchanging the order of integration, we have

$$\int_{\tilde{G}} \alpha(g') \int_{K} \sigma(k^{-1})\Phi(zkz^{-1}gg'^{-1})\, dkdg' = \int_{\tilde{G}} \alpha(g')(M_z\Phi)(gg'^{-1})\, dg' = (\alpha * M_z\Phi)(g)\,. \ \square$$

The symmetrization operator M_z also preserves analyticity.

Lemma 1.4.3. If Φ is analytic, then $M_z\Phi$ too is analytic.

Proof. Let $g_0 \in \tilde{G}$ and let $\{x_1, \ldots, x_n\}$ be a coordinate system valid in an open neighborhood V of g_0 ($n = \dim \tilde{G}$). Recall

$$(M_z\Phi)(g) = \int_K \sigma(k^{-1})\Phi(zkz^{-1}g)\, dk \quad (g \in \tilde{G})\,.$$

Since the map $G \times K \to \mathbf{C}$: $(g, k) \mapsto \Phi(zkz^{-1}g)$ is real analytic, and K is compact, there exist a finite set of coordinate neighborhoods $U_\alpha \subset K$ whose union equals K, and a neighborhood N of g_0 in V, such that the function $\Phi(zkz^{-1}g)$ is given by an absolutely and uniformly convergent power series:

$$\Phi(zk(y_1, \ldots, y_p)z^{-1}g(x_1, \ldots, x_n)) = P_\alpha(x_1, \ldots, x_n, y_1, \ldots, y_p) \qquad y \in N,\ k \in U_\alpha\,,$$

where $\{y_1, \ldots, y_p\}$ is a system of coordinates on U_α ($p = \dim K$). Now, since K is compact, there is a partition of unity $\{\phi_\alpha\}$ subordinate to the covering $\{U_\alpha\}$ of K. If one regards $P_\alpha(x_1, \ldots, x_n, y_1, \ldots, y_p)$ as a power series in $(x_1, \ldots, x_n)$ $(g(x_1, \ldots, x_n) \in N)$ with coefficient which are functions of $(y_1, \ldots, y_p)$ $(k(y_1, \ldots, y_p) \in U_\alpha)$, then $\phi_\alpha(k)P_\alpha(x_1, \ldots, x_n, y_1, \ldots, y_p)$ naturally becomes a power series in $(x_1, \ldots, x_n)$ $(g(x_1, \ldots, x_n) \in N)$ whose coefficients are functions on K: $P_\alpha(x_1, \ldots, x_n, k)$. Moreover, it is clear that

$$\Phi(zkz^{-1}g(x_1, \ldots, x_n)) = \sum_\alpha \phi_\alpha(k)P_\alpha(x_1, \ldots, x_n, k)\,.$$

Therefore, $\sigma(k^{-1})\Phi(zkz^{-1}g)$ is a power series in $(x_1, \ldots, x_n)$ that converges uniformly on $N \times K$.

We can then integrate term by term to obtain a power series representation for $(M_z\Phi)(g)$ on N.

□

The next proposition gives the crucial property of the symmetrization operator.

Proposition 1.4.2. Let Φ be a smooth σ-function on $\tilde{G}$. For any $D \in \boldsymbol{D}(\tilde{G})$, we have

$$[D(M_z\Phi)](z) = [D_K(M_z\Phi)](z)$$

where

$$D_K = \int_K \mathrm{Ad}(k')D\, dk' .$$

Proof. By definition

$$[D_K(M_z\Phi)](z) = \int_K [D(M_z\Phi)^{k'}](zk'^{-1})\, dk' ;$$

and

$$(M_z\Phi)^{k'}(g) = (M_z\Phi)(gk') = \int_K \sigma(k^{-1})\Phi(zkz^{-1}gk')\, dk$$

$$= \int_K \sigma(k^{-1})\sigma(k')\Phi(zkz^{-1}g)\, dk = (M_z\sigma(k')\Phi)(g) ,$$

so

$$(M_z\Phi)^{k'} = M_z\sigma(k')\Phi .$$

Then

$$[D_K(M_z\Phi)](z) = \int_K [D(M_z\sigma(k')\Phi)](zk'^{-1})\, dk'$$

$$= \int_K [D(M_z\sigma(k')\Phi)](zk'^{-1}z^{-1}z)\, dk'$$

$$= \int_K [D^{\,zk'^{-1}z^{-1}}(M_z\sigma(k')\Phi)](z)\, dk' .$$

Now

$$
{}^{zk'^{-1}z^{-1}}(M_z\sigma(k')\Phi)(g) = (M_z\sigma(k')\Phi)(zk'^{-1}z^{-1}g)
$$

$$
= \int_K \sigma(k^{-1})(\sigma(k')\Phi)(zkz^{-1}zk'^{-1}z^{-1}g)\, dk
$$

$$
= \int_K \sigma(k^{-1})\sigma(k')\Phi(zkk'^{-1}z^{-1}g)\, dk\,,
$$

under $kk'^{-1} \mapsto k$,

$$
= \int_K \sigma(k^{-1})\sigma(k'^{-1})\sigma(k')\Phi(zkz^{-1}g)\, dk
$$

$$
= \int_K \sigma(k^{-1})\Phi(zkz^{-1}g)\, dk = (M_z\Phi)(g)\,.
$$

Hence

$$
{}^{zk'^{-1}z^{-1}}(M_z\sigma(k')\Phi) = M_z\Phi\,.
$$

Therefore,

$$
[D_K(M_z\Phi)](z) = \int_K [D(M_z\Phi)](z)\, dk' = [D(M_z\Phi)](z)\,. \quad \square
$$

We can now proceed to deduce that

$$
\alpha * \Phi = \hat{\alpha}_\Phi \cdot \Phi
$$

for some constant function $\hat{\alpha}_\Phi$ on $\tilde{G}$. Let $z_0 \in \tilde{G}$ be such that $\Phi(z_0) \neq 0$. Since $(M_{z_0}\Phi)(z_0) = \Phi(z_0) \neq 0$, and the function $(z, g) \mapsto (M_z\Phi)(g)$ is continuous, there is a neighborhood U of z_0 such that $(M_z\Phi)(g)$ is nonzero on $U \times U$. Note that $\Phi(z) = (M_z\Phi)(z)$ is nonzero on U. Because both Φ and $\alpha * \Phi$ are analytic σ-functions, it suffices to show that

$$
\frac{(\alpha * \Phi)(z)}{\Phi(z)}
$$

is constant on U. This is a corollary to Theorem 1.4.3.

Theorem 1.4.3. Given $z \in U$, if $z' \in \tilde{G}$ is such that $z'z \in U$, then

$$\frac{[M_{z'z}\Phi](z'g)}{(M_{z'z}\Phi)(z'z)} = \frac{[M_z\Phi](g)}{(M_z\Phi)(z)} \qquad (g \in \tilde{G}) \tag{1.4.2}$$

and

$$\frac{[M_{z'z}(\alpha * \Phi)](z'g)}{(M_{z'z}\Phi)(z'z)} = \frac{[M_z(\alpha * \Phi)](g)}{(M_z\Phi)(z)} \qquad (g \in \tilde{G})\,. \tag{1.4.3}$$

Proof. We prove the equation (1.4.2) first. Starting with the definition of $M_{z'z}$,

$$[M_{z'z}\Phi](z'g) = \int_K \sigma(k^{-1})\Phi(z'zkz^{-1}z'^{-1}z'g)\,dk$$

$$= \int_K \sigma(k^{-1})\Phi(z'zkz^{-1}g)\,dk = [M_z\,{}^{z'}\Phi](g) \qquad (g \in \tilde{G})\,.$$

For $D \in \boldsymbol{D}_K(\tilde{G})$,

$$D(M_z\,{}^{z'}\Phi) = M_z D\,{}^{z'}\Phi$$

$$= M_z\,\chi(D)\,{}^{z'}\Phi$$

$$= \chi(D) M_z\,{}^{z'}\Phi\,.$$

For $D \in \boldsymbol{D}(\tilde{G})$, since ${}^{z'}\Phi$ is a σ-function, Proposition 1.4.2 asserts that

$$[D(M_z\,{}^{z'}\Phi)](z) = [D_K(M_z\,{}^{z'}\Phi)](z)\,,$$

where

$$D_K = \int_K \mathrm{Ad}(k)D\,dk \in \boldsymbol{D}_K(\tilde{G})\,.$$

Thus

$$[D(M_z\,{}^{z'}\Phi)](z) = \chi(D_K)(M_z\,{}^{z'}\Phi)(z)$$

for all $D \in \boldsymbol{D}(\tilde{G})$. Of course, $M_z\Phi$ has the same property. Then

$$D\Big[(M_z\Phi)(z)[M_z\,{}^{z'}\Phi]\Big](z) = D\Big[(M_z\,{}^{z'}\Phi)(z)[M_z\Phi]\Big](z)$$

for all $D \in \mathcal{D}(\tilde{G})$. Because both $M_z{}^{z'}\Phi$ and $M_z\Phi$ are analytic σ-functions,

$$(M_z\Phi)(z)[M_z{}^{z'}\Phi] = (M_z{}^{z'}\Phi)(z)[M_z\Phi]\,.$$

As

$$[M_z{}^{z'}\Phi](g) = [M_{z'z}\Phi](z'g)$$

from above, and

$$(M_z{}^{z'}\Phi)(z) = {}^{z'}\Phi(z) = \Phi(z'z) = (M_{z'z}\Phi)(z'z)\,,$$

we have the equation (1.4.2) if $z, z'z \in U$.

To prove the equation (1.4.3), observe that

$$\frac{[M_z(\alpha * \Phi)(g)}{(M_z\Phi)(z)} = \frac{[\alpha * M_z\Phi](g)}{(M_z\Phi)(z)}$$

by Lemma 1.4.2,

$$= \left[\alpha * \frac{M_z\Phi}{(M_z\Phi)(z)}\right](g) = (\alpha * \omega_z)(g)\,,$$

where

$$\omega_z = \frac{M_z\Phi}{(M_z\Phi)(z)}\,.$$

Then

$$\frac{[M_{z'z}(\alpha * \Phi)](z'g)}{(M_{z'z}\Phi)(z'z)} = (\alpha * \omega_{z'z})(z'g) = \int_{\tilde{G}} \alpha(g')\omega_{z'z}(z'gg'^{-1})\,dg'$$

$$= \int_{\tilde{G}} \alpha(g')\omega_z(gg'^{-1})\,dg' = (\alpha * \omega_z)(g)\,,$$

because

$$\omega_{z'z}(z'gg'^{-1}) = \omega_z(gg'^{-1})$$

by the equation (1.4.2), which we have just established. Thus, the equation (1.4.3) too has been proved. □

Corollary. The function $(\alpha * \Phi)(z)/\Phi(z)$ is constant on U.

Proof. Given $(z, g) \in U \times U$, if $z' \in \tilde{G}$ is such that $(z'z, z'g) \in U \times U$, then the equations (1.4.2) and (1.4.3) combined give

$$\frac{[M_{z'z}(\alpha * \Phi)](z'g)}{[M_{z'z}\Phi](z'g)} = \frac{[M_z(\alpha * \Phi)](g)}{[M_z\Phi](g)} .$$

If we let $(z, g) = (z_0, z_0) \in U \times U$, then we have

$$\frac{[M_{z'z_0}(\alpha * \Phi)](z'z_0)}{[M_{z'z_0}\Phi](z'z_0)} = \frac{[M_{z_0}(\alpha * \Phi)](z_0)}{[M_{z_0}\Phi](z_0)}$$

as long as $z'z_0 \in U$. Now since $\alpha * \Phi$ is a σ-function,

$$[M_{z'z_0}(\alpha * \Phi)](z'z_0) = (\alpha * \Phi)(z'z_0) ,$$

and

$$[M_{z_0}(\alpha * \Phi)](z_0) = (\alpha * \Phi)(z_0) .$$

Of course,

$$[M_{z'z_0}\Phi](z'z_0) = \Phi(z'z_0), \quad [M_{z_0}\Phi](z_0) = \Phi(z_0) ,$$

so

$$\frac{(\alpha * \Phi)(z'z_0)}{\Phi(z'z_0)} = \frac{(\alpha * \Phi)(z_0)}{\Phi(z_0)}$$

as long as $z'z_0 \in U$, which implies

$$\frac{(\alpha * \Phi)(z)}{\Phi(z)} = \frac{(\alpha * \Phi)(z_0)}{\Phi(z_0)} \quad \text{for all } z \in U . \quad \square$$

To complete this development, we show that $\hat{\alpha}_\Phi$ depends on Φ only to the extent of χ.

Proposition 1.4.3. Let Φ and Φ' be two smooth σ-functions on $\tilde{G}$ which are eigenfunctions for all of $\mathbf{Z}(\tilde{G})$. If

$$D\Phi = \chi(D)\Phi, \quad D\Phi' = \chi'(D)\Phi' \quad \text{for } D \in \mathbf{Z}(\tilde{G}),$$

and

$$\chi(D) = \chi'(D) \quad \text{for all } D \in \mathbf{Z}(\tilde{G}),$$

then

$$\hat{\alpha}_\Phi = \hat{\alpha}_{\Phi'} .$$

Proof. If, as before, z_0 is such that $\Phi(z_0) \neq 0$, then $\alpha * \Phi = \hat{\alpha}_\Phi \cdot \Phi$ implies

$$\frac{M_{z_0}(\alpha * \Phi)}{\Phi(z_0)} = \hat{\alpha}_\Phi \cdot \frac{M_{z_0}\Phi}{\Phi(z_0)} .$$

Since

$$M_{z_0}(\alpha * \Phi) = \alpha * M_{z_0}\Phi ,$$

and

$$\Phi(z_0) = (M_{z_0}\Phi)(z_0) ,$$

we have

$$\alpha * \frac{M_{z_0}\Phi}{(M_{z_0}\Phi)(z_0)} = \hat{\alpha}_\Phi \cdot \frac{M_{z_0}\Phi}{(M_{z_0}\Phi)(z_0)} . \tag{1.4.4}$$

Clearly $M_{z_0}\Phi/(M_{z_0}\Phi)(z_0)$ is a function of value 1 at z_0; moreover, for any $D \in \mathcal{D}(\tilde{G})$,

$$D\left[\frac{M_{z_0}\Phi}{(M_{z_0}\Phi)(z_0)}\right](z_0) = D_k\left[\frac{M_{z_0}\Phi}{(M_{z_0}\Phi)(z_0)}\right](z_0)$$

$$= \chi(D_K)\,\frac{(M_{z_0}\Phi)(z_0)}{(M_{z_0}\Phi)(z_0)} = \chi(D_K)\,,$$

where $D_K = \int_K \mathrm{Ad}(k)D\,dk$. Now, we know from Theorem 1.4.2 that

$$\chi(D_K) = \chi'(D_K) \quad \text{for all } D \in \boldsymbol{D}(\tilde{G})\,.$$

Let z_0 be such that $\Phi(z_0)$, $\Phi'(z_0) \neq 0$, then the functions

$$\frac{M_{z_0}\Phi}{(M_{z_0}\Phi)(z_0)}, \quad \frac{M_{z_0}\Phi'}{(M_{z_0}\Phi')(z_0)}$$

are identical by Taylor's formula. So the equation (1.4.4) implies

$$\hat{\alpha}_\Phi = \hat{\alpha}_{\Phi'}\,. \quad \square$$

From the equation (1.4.1), it now follows that

$$\alpha * \Phi_\Lambda = \hat{\alpha}^o(\Lambda)\Phi_\Lambda\,,$$

where $\hat{\alpha}^o(\Lambda) = \hat{\alpha}_\Lambda$, which is the constant function in

$$\alpha_\Lambda * \Phi = \hat{\alpha}_\Lambda\Phi\,.$$

It is clear that for a given σ and a given χ, $\hat{\alpha}^o(\Lambda)$ does not depend on Φ.

Hereafter the identity element of G will be denoted by e. The remainder of this section is devoted to showing that $\hat{\alpha}^o$ is either zero or nonconstant.

Proposition 1.4.4. The function $\hat{\alpha}^o$ is holomorphic on $\boldsymbol{a}_{\mathbb{C}}^*$. Moreover, if $\hat{\alpha}^o(\Lambda)$ is not identically zero, it is nonconstant.

Proof. We have

$$\alpha_\Lambda * \Phi = \hat{\alpha}_\Lambda\Phi = \hat{\alpha}^o(\Lambda)\Phi\,.$$

Recall that

$$(\alpha_\Lambda * \Phi)(m) = \int_M \int_A \int_N \int_K \alpha(n'a'm'^{-1}mk'^{-1})\sigma(k')e^{(\Lambda+\rho)(\log a')}\Phi(m')\, dk'dn'da'dm' \qquad (m \in M).$$

It is evident that for every $m \in M$, $(\alpha_\Lambda * \Phi)(m)$ is holomorphic in Λ for $\Lambda \in \boldsymbol{a}_C^*$. Let $m \in M$ be such that $\Phi(m) \neq 0$, then

$$\hat{\alpha}^o(\Lambda) = \frac{(\alpha_\Lambda * \Phi)(m)}{\Phi(m)},$$

which is holomorphic in Λ. Upon changing the order of integration, $(\alpha_\Lambda * \Phi)(m)$ becomes

$$\int_A \left[\int_M \int_N \int_K \alpha(n'a'm'^{-1}mk'^{-1})\sigma(k')\Phi(m')\, dk'dn'dm'\right] e^{(\Lambda+\rho)(\log a')}\, da' \qquad (m \in M),$$

where the triple inner integral can be viewed as a convolution $\alpha_{a'} * \Phi$, with

$$\alpha_{a'}(m) = \int_N \int_K \alpha(n'a'mk'^{-1})\sigma(k')\, dk'dn' \qquad (m \in M).$$

One verifies as in the proof of Proposition 1.4.1 that $\alpha_{a'} * \Phi$ too is a σ_M-function. Then the fundamental principle asserts that

$$\alpha_{a'} * \Phi = \hat{\alpha}_{a'}\Phi$$

for some constant function $\hat{\alpha}_{a'}$ on M.

For a given $m \in M$, one easily sees that $(\alpha_{a'} * \Phi)(m)$ is a continuous compactly supported function on A (cf. Lemma 1.4.1), so that $\hat{\alpha}_{a'}$ is also a continuous compactly supported function on A.

We now have

$$\begin{aligned}(\alpha_\Lambda * \Phi)(m) &= \int_A \hat{\alpha}_{a'}\Phi(m)e^{(\Lambda+\rho)(\log a')}\, da' \\ &= \left[\int_A \hat{\alpha}_{a'}e^{(\Lambda+\rho)(\log a')}\, da'\right]\Phi(m),\end{aligned}$$

so that

$$\hat{\alpha}^{o}(\Lambda) = \int_A \hat{\alpha}_{a'} e^{(\Lambda+\rho)(\log a')}\, da' \qquad (\Lambda \in \boldsymbol{a}_{\mathbf{C}}^{*}) .$$

Fix some nonzero $\tau \in \boldsymbol{a}^*$ and let $\Lambda(t) = -\rho - it\tau$ ($t \in \mathbf{R}$). Then

$$\hat{\alpha}^{o}(\Lambda(t)) = \int_A \hat{\alpha}_{a'} e^{-it\tau(\log a')}\, da' \qquad (t \in \mathbf{R}) .$$

We will show that

$$\lim_{t \to \infty} \hat{\alpha}^{o}(\Lambda(t)) = 0 .$$

(This is similar to the classical Riemann-Lebesgue Lemma.) The result would contradict any claim that $\hat{\alpha}^{o}$ is a nonzero constant function on $\boldsymbol{a}_{\mathbf{C}}^{*}$. Pick a basis of $\boldsymbol{a}^*$ and represent τ as $(\lambda_1, \ldots, \lambda_k, \ldots, \lambda_n)$ where $\lambda_k \neq 0$. Use the dual basis for $\boldsymbol{a}$ and set $H_t = (0, \ldots, \frac{\pi}{\lambda_k t}, \ldots, 0)$ $(t > 0)$. Then $\tau(H_t) = \frac{\pi}{t}$ and $\lim_{t \to \infty} H_t = 0$. Let $a(t) = \exp H_t$. Obviously, $\lim_{t \to \infty} a(t) = e$.

We now make a change of variable in the integral expression for $\hat{\alpha}^{o}(\Lambda(t))$. Thus under $a' \mapsto a(t)a'$,

$$\begin{aligned}
\hat{\alpha}^{o}(\Lambda(t)) &= \int_A \hat{\alpha}_{a(t)a'} e^{-it\tau(\log a_t a')}\, da' \\
&= \int_A \hat{\alpha}_{a(t)a'} e^{-it\tau(\log a_t)} e^{-it\tau(\log a')}\, da' \\
&= \int_A \hat{\alpha}_{a(t)a'} e^{-i\pi} e^{-it\tau(\log a')}\, da' \\
&= -\int_A \hat{\alpha}_{a(t)a'} e^{-it\tau(\log a')}\, da' .
\end{aligned}$$

Hence

$$2\hat{\alpha}^{o}(\Lambda(t)) = \int_A (\hat{\alpha}_{a'} - \hat{\alpha}_{a(t)a'}) e^{-it\tau(\log a')}\, da' .$$

It is clear that

$$\lim_{t \to \infty} \hat{\alpha}_{a'} \quad \hat{}_{a(t)a'} = 0 .$$

We intend to apply dominated convergence. Note that

$$|e^{-it\tau(\log a')}| = 1 .$$

Since $\hat{\alpha}_{a'}$ is a continuous compactly supported function, it is bounded. Then $\hat{\alpha}_{a(t)a'}$ too are bounded, uniformly with respect to t. Moreover, if W denotes the support of $\hat{\alpha}_{a'}$, then

$$\hat{\alpha}_{a(t)a'} \neq 0 \Rightarrow a(t)a' \in W \Rightarrow a' \in a(t)^{-1}W .$$

For all large t, $a(t)^{-1}$ belongs to a relatively compact neighborhood of e, so $a' \in \tilde{W} \supset W$ for some compact set $\tilde{W}$, and

$$2\hat{\alpha}^{o}(\Lambda(t)) = \int_{\tilde{W}} (\hat{\alpha}_{a'} - \hat{\alpha}_{a(t)a'})e^{-it\tau(\log a')}\, da' .$$

The dominated convergence theorem now asserts that

$$\lim_{\to \infty} 2\hat{\alpha}^{o}(\Lambda(t)) = 0 . \quad \square$$

CHAPTER 2

THE COMPACT OPERATORS

2.1 Introduction

If α is a smooth compactly supported function on G such that

$$\alpha(k^{-1}gk) = \alpha(g) \qquad \text{for all } g \in G, k = K ,$$

then

$$\alpha * E = \hat{\alpha}^{o}(\Lambda)E ,$$

which we rewrite as

$$(2.1.1) \qquad (\alpha * \ - \hat{\alpha}^{o}(\Lambda))E = 0 .$$

The Selberg transform $\hat{\alpha}^{0}(\Lambda)$ will play a role in describing the set of poles of the meromorphic continuation of the Eisenstein series. It is either identically zero or it is a nonconstant entire function. Furthermore, for every $\Lambda \in \boldsymbol{a}_{C}^{*}$ there exists an α such that $\hat{\alpha}^{o}(\Lambda) \neq 0$ (Theorem 2.2.1). In fact, α will be compactly supported in a neighborhood of the identity element which, of course, depends on Λ. The compact operator $\mathbf{K}_{\alpha}$ will be defined with a kernel function α of this type.

For $f \in L^{2}(\Gamma \setminus G)$, let f^{r} be the constant term of f with respect to the rth maximal standard cuspidal subgroup. Along with the equation (2.1.1), Godement's two estimates motivate the definition of $\mathbf{K}_{\alpha}$. The first one says, in very rough terms, that for $f \in L^{2}(\Gamma \setminus G)$, $\alpha * f(g) - \alpha * f^{r}(g)$ is bounded in the direction of the rth simple root. The second estimate is a similar one for slowly increasing Γ-automorphic functions. To utilize Godement's estimates fully, we shall introduce certain truncation factors δ_{r} (Lemma 2.4.2). The operator $\mathbf{K}_{\alpha}$ is then defined to be $\alpha * \alpha * \mathbf{K}_{\alpha}^{o}$, where

$$\mathbf{K}_\alpha^o f = \left[\alpha * f - \sum_{r=1}^{m} \delta_r(\alpha * f^r) \right]_F .$$

Then it can be deduced at once that (Proposition 2.4.2) there exists $\kappa > 0$ such that for all $f \in L^2(\Gamma \backslash G)$,

$$|\mathbf{K}_\alpha f(g)| \leq \kappa \|f\|_2 \qquad \text{for all } g \in G .$$

From this the compactness of $\mathbf{K}_\alpha$ can be deduced in the usual fashion (Lemma 2.4.3, Theorem 2.4.1).

Later in Chapter 3, $\mathbf{K}_\alpha$ will be applied to the Eisenstein series. In view of the equation (2.1.1), this will bring in a set Fredholm equations. The 2-fold convolution $\alpha * \alpha *$ in the definition of $\mathbf{K}_\alpha$ is largely to ensure the smoothness of the equation for the Eisenstein series and of the Fredholm equations.

2.2. Construction of Kernel

Our objective in this section is the construction of a smooth compactly supported function α on G such that

$$\alpha(k^{-1}gk) = \alpha(g) \qquad \text{for all } g \in G, k \in K .$$

Then, as shown in Section 1.4, we have

$$\int_G \alpha(h^{-1}g)\Phi_\Lambda(h)\, dh = \hat{\alpha}^o(\Lambda)\Phi_\Lambda(g) .$$

We shall actually prove that for every $\Lambda \in \boldsymbol{a}_C^*$, there is a smooth compactly supported function α on G such that

$$\alpha(k^{-1}gk) = \alpha(g) \qquad \text{for all } g \in G, k \in K ,$$

and

$$\hat{\alpha}^{o}(\Lambda) \neq 0 .$$

Proposition 2.1.1. For every smooth compactly supported function μ on G, the integral

$$\alpha(g) = \int_K \mu(k^{-1}gk)\, dk \qquad (g \in G)$$

defines a smooth compactly supported function α on G such that

$$\alpha(k^{-1}gk) = \alpha(g) \qquad \text{for all } g \in G, k \in K .$$

Proof. Since K is compact, α is clearly well defined. The smoothness of α can be verified in a straightforward manner. For $k' \in K$,

$$\alpha(k'^{-1}gk) = \int_K \mu(k^{-1}k'^{-1}gk'k)\, dk = \int_K \mu((k'k)^{-1}g(k'k))\, dk .$$

As $k'k \mapsto k$, the integral becomes

$$\int_K \mu(k^{-1}gk)\, dk = \alpha(g) .$$

So α has the desired invariance property. Denote by C the compact support of μ. For $k \in K$, $\mu(k^{-1}gk) \neq 0$ implies that $k^{-1}gk \in C$. Then $g \in kCk^{-1} \subset KCK^{-1}$. Since KCK^{-1} is compact, α is compactly supported. □

Theorem 2.2.1. For every $\Lambda \in \mathfrak{a}_{\mathbb{C}}^{*}$, there is a smooth compactly supported function α on G such that

$$\alpha(k^{-1}gk) = \alpha(g) \qquad \text{for all } g \in G, k \in K ,$$

and

$$\hat{\alpha}^{o}(\Lambda) \neq 0 .$$

Proof. Given $\Lambda \in \mathfrak{a}_C^*$, fix $g_0 = namk \in G$ such that

$$\Phi_\Lambda(g_0) = \sigma(k)\Phi(m)e^{(-\Lambda-\rho)(\log a)} \neq 0 .$$

Obviously, it suffices to find $m \in M$ such that $\Phi(m) \neq 0$. Now the continuity of Φ_Λ implies the existence of a neighborhood W of g_0 in which

$$|\Phi_\Lambda(h) - \Phi_\Lambda(g_0)| < |\Phi_\Lambda(g_0)| \qquad (h \in W) .$$

We claim that if a smooth compactly supported function α satisfies

$$\alpha(k^{-1}gk) = \alpha(g) \qquad \text{for all } g \in G, k \in K ,$$

as well as the properties: α is nonnegative, $\alpha(h^{-1}g_0) = 0$ for all $h \in W^c$, and

$$\int_G \alpha(h^{-1}g_0)\, dh = 1 ,$$

then $\hat{\alpha}^o(\Lambda) \neq 0$. To see this, note first that the last property implies

$$\int_G \alpha(h^{-1}g_0)\Phi_\Lambda(h)\, dh - \Phi_\Lambda(g_0) = \int_G \alpha(h^{-1}g_0)(\Phi_\Lambda(h) - \Phi_\Lambda(g_0))\, dh$$

$$= \int_W \alpha(h^{-1}g_0)(\Phi_\Lambda(h) - \Phi_\Lambda(g_0))\, dh$$

$$+ \int_{W^c} \alpha(h^{-1}g_0)(\Phi_\Lambda(h) - \Phi_\Lambda(g_0))\, dh .$$

Since $\alpha(h^{-1}g_0) = 0$ for $h \in W^c$, this sum is simply

$$\int_W \alpha(h^{-1}g_0)(\Phi_\Lambda(h) - \Phi_\Lambda(g_0))\, dh .$$

Therefore,

$$\left|\int_G \alpha(h^{-1}g_0)\Phi_\Lambda(h)\, dh - \Phi_\Lambda(g_0)\right| \leq \int_W |\alpha(h^{-1}g_0)|\; |\Phi_\Lambda(h) - \Phi_\Lambda(g_0)|\, dh$$

$$< |\Phi_\Lambda(g_0)| \int_W |\alpha(h^{-1}g_0)|\, dh .$$

As α is nonnegative, the last quantity equals

$$|\Phi_\Lambda(g_0)| \int_W \alpha(h^{-1}g_0)\,dh = |\Phi_\Lambda(g_0)| \;.$$

On the other hand,

$$\Big|\int_G \alpha(h^{-1}g_0)\Phi_\Lambda(h)\,dh \; - \; \Phi_\Lambda(g_0)\Big| = |\hat{\alpha}^o(\Lambda)\Phi_\Lambda(g_0) - \Phi_\Lambda(g_0)|$$

$$= |\Phi_\Lambda(g_0)|\,|\hat{\alpha}^o(\Lambda) - 1| \;.$$

Hence

$$|\Phi_\Lambda(g_0)|\,|\hat{\alpha}^o(\Lambda) - 1| < |\Phi_\Lambda(g_0)| \;,$$

which implies

$$|\hat{\alpha}^o(\Lambda) - 1| < 1 \;,$$

and therefore,

$$\hat{\alpha}^0(\Lambda) \neq 0 \;.$$

We now proceed to construct α. Refer to the formula for α in Proposition 2.2.1. As usual e denotes the identity element of G. Suppose there exists a neighborhood V of e such that

$$V \subset \Big(\bigcup_{k \in K} k^{-1}(W^c)^{-1} g_0 k \Big)^c \;,$$

then if μ is a smooth nonnegative function compactly supported in V with $\mu(e) \neq 0$, we have for $h \in W^c$,

$$\alpha(h^{-1}g_0) = \int_K \mu(k^{-1}h^{-1}g_0 k)\,dk = 0 \;.$$

Furthermore,

$$\alpha(e) \;=\; \mu(e)\int_K \,dk \;\neq\; 0 \;,$$

hence the integral

$$\int_G \alpha(h^{-1}g_0)\, dh$$

is nonzero and α can be normalized by dividing by this integral. The nonnegativity of α is clear.

As to the existence of V, it suffices to show that e is not a limit point of

$$\bigcup_{k \in K} k^{-1}(W^c)^{-1}g_0k\,.$$

Suppose the contrary. There exists a sequence $\{k_i^{-1}x_ig_0k_i\}$, where $x_i \in (W^c)^{-1} = (W^{-1})^c$, that converges to e. Recall that G is a submanifold of GL(n, **R**), which has the topology of a normed linear space. Then

$$k_i^{-1}x_ig_0k_i \to e \quad \text{if and only if} \quad \| k_i^{-1}x_ig_0k_i - e \| \to 0\,.$$

Now, since K is compact,

$$\| k_i \| \, \| k_i^{-1}x_ig_0k_i - e \| \, \| k_i^{-1} \| \leq M^2 \| k_i^{-1}x_ig_0k_i - e \|$$

for some number M, where $\| k \| \leq M$ for all $k \in K$. Hence

$$\| k_i \| \, \| k_i^{-1}x_ig_0k_i - e \| \, \| k_i^{-1} \| \to 0\,.$$

Thus

$$\| k_i \| \, \| k_i^{-1}x_ig_0k_i - e \| \, \| k_i^{-1} \| \geq \| k_i(k_i^{-1}x_ig_0k_i - e)k_i^{-1} \| = \| x_ig_0 - e \|$$

implies

$$\| x_ig_0 - e \| \to 0\,.$$

Then

$$\| x_ig_0 - e \| \, \| g_0^{-1} \| \to 0\,.$$

But

$$\| x_ig_0 - e \| \, \| g_0^{-1} \| \geq \| x_i - g_0^{-1} \|\,,$$

so $x_i \to g_0^{-1}$. We have arrived at a contradiction, because $g_0^{-1} \in W^{-1}$, which is open, whereas

$x_i \in (W^{-1})^c$. Therefore, e is not a limit point of $\bigcup_{k \in K} k^{-1}(W^c)^{-1} g_0 k$. □

We close this section by deducing an easy corollary of Theorem 2.2.1. Suppose f is a σ-function for some character σ of K, and is an eigenfunction for all of $\mathbf{Z}(G)$. If, for some $\Lambda \in \mathfrak{a}_C^*$, f has the same set of eigenvalues as Φ_Λ for some $\Phi \in {}^0L^2(\Gamma_M \backslash M, \sigma_M, \chi)$, χ a character of $\mathbf{Z}(M)$, then there exists a smooth compactly supported function α,

$$\alpha(k^{-1}gk) = \alpha(g) \qquad \text{for all } g \in G, k \in K,$$

such that

$$\alpha * f = f.$$

A far more general phenomenon underlies this fact. The following result is due to Harish-Chandra (Harish-Chandra [4]; also Baily [1]).

Theorem 2.2.2. Suppose f is a K-finite and $\mathbf{Z}(G)$-finite function on G. There exists a smooth compactly supported function α,

$$\alpha(k^{-1}gk) = \alpha(g) \qquad \text{for all } g \in G, k \in K,$$

such that

$$\alpha * f = f.$$

2.3. Convolutions on $L^2(\Gamma \backslash G)$

Before the introduction of the compact operators, we state and prove in this section some basic facts about convolution on $L^2(\Gamma \backslash G)$. We begin by quoting a result from reduction theory. Recall that each element of G has its usual matrix norm.

Lemma 2.3.1. Let α be a bounded compactly supported function on G. We can choose numbers C, N > 0 such that

$$\sum_{\gamma \in \Gamma} |\alpha(x\gamma y)| \leq C \|y\|^N \qquad \text{for all } x, y \in G.$$

(See Harish-Chandra [2, p. 9].)

Lemma 2.3.1 says in particular that for any compact subset Ω of G, $\Gamma \cap \Omega$ is a finite set. To see this, let α be the characteristic function of Ω, and x, y = e.

In the definition of the compact operator, the kernel function will be a smooth compactly supported function on G. Henceforth, unless otherwise stated, α will be assumed to be smooth.

Recall that for $f \in L^2(\Gamma \backslash G)$,

$$f^r(g) = \int_{N_r \cap \Gamma \backslash N_r} f(ng)\, dn \qquad (g \in G),$$

where N_r is as in $P_r = N_r A_r M_r$, P_r being the rth maximal standard cuspidal subgroup, $r = 1, \ldots, m$. The next lemma says in particular that $\alpha * f$ and $\alpha * f^r$ are well defined. For $f \in L^2(\Gamma \backslash G)$, as usual $\|f\|_2$ denotes the L^2-*norm* of f.

Lemma 2.3.2. There exist constants N, B > 0 such that for all $f \in L^2(\Gamma \backslash G)$,

$$|\alpha * f(g)|, \ |\alpha * f^r(g)| \leq B \|g\|^N \|f\|_2 \qquad (g \in G), \quad r = 1, \ldots, m.$$

Proof. Let F be a fundamental domain for $\Gamma \backslash G$ and observe that

$$\begin{aligned}\int_G |\alpha(h^{-1}g) f(h)|\, dh &= \sum_{\gamma \in \Gamma} \int_F |\alpha((\gamma h)^{-1} g)| \ |f(\gamma h)|\, dh \\ &= \sum_{\gamma \in \Gamma} \int_F |\alpha((\gamma h)^{-1} g)| \ |f(g)|\, dh \\ &= \int_F \sum_{\gamma \in \Gamma} |\alpha((\gamma h)^{-1} g)| \cdot |f(h)|\, dh\end{aligned}$$

$$= \int_{\Gamma\backslash G} \sum_{\gamma\in\Gamma} |\alpha(h^{-1}\gamma g)|\cdot|f(h)|\,dh\,.$$

By Lemma 2.3.1, the last quantity is less than or equal to

$$C\,\|g\|^N \mathrm{Vol}(\Gamma\backslash G)^{1/2}\,\|f\|^2 < \infty\,.$$

We also have

$$\int_G |\alpha(h^{-1}g)f^r(h)|\,dh = \int_G \Big|\alpha(h^{-1}g)\int_{N_r\cap\Gamma\backslash N_r} f(nh)\,dn\Big|\,dh$$

$$\leq \int_G \left[\int_{N_r\cap\Gamma\backslash N_r} |\alpha(h^{-1}g)f(nh)|\,dn\right] dh$$

$$= \int_G \left[\int_{F^o} |\alpha(h^{-1}g)f(nh)|\,dn\right] dh\,,$$

where $F^o \subset N_r$ is a fundamental domain for $N_r \cap \Gamma \backslash N_r$. On the other hand, under the change of variable $nh \mapsto h$,

$$\int_G |\alpha(h^{-1}g)f(nh)|\,dh = \int_G |\alpha((n^{-1}h)^{-1}g)f(h)|\,dh$$

$$= \int_G |\alpha(h^{-1}(ng))f(h)|\,dh$$

$$\leq C\,\|ng\|^N \mathrm{Vol}(\Gamma\backslash G)^{1/2}\,\|f\|_2$$

$$= C\,\|n\|^N\,\|g\|^N \mathrm{Vol}(\Gamma\backslash G)^{1/2}\,\|f\|_2$$

as above. Therefore, since F^o is relatively compact,

$$\int_{F^o} \left[\int_G |\alpha(h^{-1}g)f(nh)|\,dh\right] dn < \infty\,.$$

Fubini's theorem implies that the two iterated integrals are equal. The lemma's assertion is now clear. □

Next we see that convolution preserves automorphicity.

Proposition 2.3.1. If $f \in L^2(\Gamma \backslash G)$, then the convolution $\alpha * f$ is automorphic.

Proof. For $\gamma \in \Gamma$,

$$\alpha * f(\gamma g) = \int_G \alpha(h^{-1}\gamma g)f(h)\, dh = \int_G \alpha((\gamma^{-1}h)^{-1}g)f(h)\, dh$$

$$= \int_G \alpha(h^{-1}g)f(\gamma h)\, dh = \int_G \alpha(h^{-1}g)f(h)\, dh = \alpha * f(g)\, ,$$

where the third integral was derived from the second by the change of variable $\gamma^{-1}h \mapsto h$. □

Convolution also preserves boundedness.

Proposition 2.3.2. Let f be a bounded function in $L^2(\Gamma \backslash G)$, i.e., there is an $N > 0$ such that $|f(g)| \le N$ for all $g \in G$. Then there exists $B > 0$, which depends only on α, such that $|\alpha * f(g)| \le NB$ for all $g \in G$.

Proof. From the definition of convolution we have

$$\alpha * f(g) = \int_G \alpha(h^{-1}g)f(h)\, dh = \int_G \alpha((g^{-1}h)^{-1})f(h)\, dh\, .$$

Under the change of variable $g^{-1}h \mapsto h$, the last integral becomes

$$\int_G \alpha(h^{-1})f(gh)\, dh\, .$$

If we denote the compact support of α by Ω, then

$$|\alpha * f(g)| \le N \int_{\Omega^{-1}} |\alpha(h^{-1})|\, dh \le NB \qquad \text{for all } g \in G\, ,$$

where $B > 0$ depends only on α. □

If instead of function in $L^2(\Gamma \backslash G)$ we work with functions in $C(\Gamma \backslash G)$, then $\alpha * f$ and $\alpha * f^r$ are also well defined. (It can easily be verified that f^r, $r = 1, \ldots, m$, are continuous functions on G.) Moreover, Propositions 2.3.1 and 2.3.2 will remain valid.

Convolution on $L^2(\Gamma \backslash G)$ can be said to have smoothing effect; the next two propositions give indication of this.

Proposition 2.3.3. Let α be a continuous compactly supported function on G. If $f \in L^2(\Gamma \backslash G)$, then $\alpha * f$ and $\alpha * f^r$ $(r = 1, \ldots, n)$ are continuous functions on G.

Proof. Let $g \in G$ be fixed. Let W be a relatively compact neighborhood of e, the identity element of G. If g' is an element of gW, $g^{-1}g' \in W$. There exists $M > 0$ such that $\|g\|, \|g'\| < M$ for all $g' \in W$. Denote the compact support of α by Ω. Since $\Omega\overline{W}$ is compact, α is uniformly continuous on $\Omega\overline{W}$, i.e., for every $\varepsilon > 0$ there exists a neighborhood U of e such that whenever $g_1, g_2 \in \Omega\overline{W}$ and $g_1^{-1}g_2 \in U$, then $|\alpha(g_1) - \alpha(g_2)| < \varepsilon$. Now

$$
\begin{aligned}
|\alpha * f(g) - \alpha * f(g')| &= \left| \int_G (\alpha(h^1 g) - \alpha(h^{-1}g'))f(h)\, dh \right| \\
&\le \int_G |\alpha(h^{-1}g) - \alpha(h^{-1}g')|\, |f(h)|\, dh \\
&= \int_{\Gamma \backslash G} \sum_{\gamma \in \Gamma} |\alpha(h^{-1}\gamma g) - \alpha(h^{-1}\gamma g')| \cdot |f(h)|\, dh\,.
\end{aligned}
$$

Write $h^{-1}\gamma g' = h^{-1}\gamma g(g^{-1}g')$, then $h^{-1}\gamma g \in \Omega$ implies $h^{-1}\gamma g(g^{-1}g') \in \Omega\overline{W}$. Either both $h^{-1}\gamma g$ and $h^{-1}\gamma g'$ belong to Ω, or both are outside Ω. If both are outside Ω,

$$
|\alpha(h^{-1}\gamma g) - \alpha(h^{-1}\gamma g')| = 0 - 0 = 0\,.
$$

So, for the purpose of estimating

$$
\sum_{\gamma \in \Gamma} |\alpha(h^{-1}\gamma g) - \alpha(h^{-1}\gamma g')|\,,
$$

we can assume that both $h^{-1}\gamma g$ and $h^{-1}\gamma g'$ belong to $\Omega\overline{W}$. Then by uniform continuity, there exists a neighborhood U of e such that whenever $(h^{-1}\gamma g)^{-1}(h^{-1}\gamma g') = g^{-1}g' \in U$, then

$$|\alpha(h^{-1}\gamma g) - \alpha(h^{-1}\gamma g')| < \varepsilon \text{ for all } h, \gamma.$$

We now further restrict g' to be in $gW \cap gU = g(W \cap U)$. Denote by χ_Ω the characteristic function of Ω. Then

$$\sum_\gamma |\alpha(h^{-1}\gamma g) - \alpha(h^{-1}\gamma g')|$$

$$\leq \Big(\sum_{\gamma\in\Gamma} \chi_\Omega(h^{-1}\gamma g) + \sum_{\gamma\in\Gamma} \chi_\Omega(h^{-1}\gamma g') \Big)\varepsilon,$$

because

$$\sum_{\gamma\in\Gamma} \chi_\Omega(h^{-1}\gamma g) + \sum_{\gamma\in\Gamma} \chi_\Omega(h^{-1}\gamma g')$$

is the maximum number of elements of Γ at which one of $\alpha(h^{-1}\gamma g)$, $\alpha(h^{-1}\gamma g')$ is nonzero. By Lemma 2.3.1,

$$\sum_{\gamma\in\Gamma} \chi_\Omega(h^{-1}\gamma g) \leq C\,\|g\|^N \quad\text{and}\quad \sum_{\gamma\in\Gamma} \chi_\Omega(h^{-1}\gamma g') \leq C\,\|g'\|^N.$$

Therefore,

$$\sum_{\gamma\in G} |\alpha(h^{-1}\gamma g) - \alpha(h^{-1}\gamma g')| \leq C(\|g\|^N + \|g'\|^N)\varepsilon \leq 2CM^N\varepsilon.$$

Then

$$|\alpha * f(g) - \alpha * f(g')| \leq \int_G |\alpha(h^{-1}g) - \alpha(h^{-1}g')|\,|f(h)|\,dh$$

$$\leq 2CM^N\varepsilon \int_{\Gamma\backslash G} |f(h)|\,dh$$

$$\leq (2CM^N \mathrm{Vol}(\Gamma\backslash G)^{1/2}\,\|f\|_2)\varepsilon.$$

It is now clear that for every $\varepsilon > 0$ there exists a neighborhood U of g such that $g' \in U$ implies $|\alpha * f(g) - \alpha * f(g')| < \varepsilon$. This proves the continuity of $\alpha * f$.

Next,

$$|\alpha * f^r(g) - \alpha * f^r(g')| = |\int_G (\alpha(h^{-1}g) - \alpha(h^{-1}g')) \int_{N_r \cap \Gamma \setminus N_r} f(nh)\, dndh|$$

$$\leq \int_{N_r \cap \Gamma \setminus N_r} \int_G |\alpha(h^{-1}g) - \alpha(h^{-1}g')|\, |f(nh)|\, dhdn$$

$$= \int_{N_r \cap \Gamma \setminus N_r} \int_G |\alpha(h^{-1}ng) - \alpha(h^{-1}ng')|\, |f(h)|\, dhdn$$

$$= \int_{N_r \cap \Gamma \setminus N_r} \int_{\Gamma \setminus G} \sum_{\gamma \in \Gamma} |\alpha(h^{-1}\gamma ng) - \alpha(h^{-1}\gamma ng')| \cdot |f(h)|\, dhdn .$$

Since $N_r \cap \Gamma \setminus N_r$ is compact, the continuity of $\alpha * f^r$ proceeds along the same line. □

Proposition 2.3.4. Let α be a smooth compactly supported function on G and $f \in L^2(\Gamma \setminus G)$. For every $X \in \boldsymbol{g}$, the Lie algebra of G,

$$X(\alpha * f) = (X\alpha) * f, \quad X(\alpha * f^r) = (X\alpha) * f^r \qquad (r = 1,\ldots, m) .$$

Proof. Let $X \in \boldsymbol{g}$. By definition, for a given $g \in G$,

$$X(\alpha * f)(g) = \lim_{t \to 0} (\alpha * f(g \exp tX) - \alpha * f(g))/t$$

$$= \lim_{t \to 0} \int_G (\alpha(h^{-1}g \exp tX) - \alpha(h^{-1}g))/t \;\; f(h)\, dh$$

$$= \lim_{t \to 0} \int_G \left[\frac{d}{dt'} \alpha(h^{-1}g \exp t'X) \right]_{t'=t_0} f(h)\, dh$$

by the mean value theorem of calculus, where $|t_0| \leq |t|$ and t_0 depends on h and t. Then

$$X(\alpha * f)(g) = \lim_{t \to 0} \int_G (X\alpha)(h^{-1}g \exp t_0X) f(h)\, dh$$

$$= \lim_{t \to 0} \int_{\Gamma \setminus G} \sum_{\gamma \in \Gamma} (X\alpha)(h^{-1}\gamma g \exp t_0X) \cdot f(h)\, dh .$$

Let δ be a positive number. We restrict t so that $|t| < \delta$. Denote the compact support of α by

Ω. Then the support of $X\alpha$ is contained in Ω. Now

$$h^{-1}\gamma g \exp t_0 X \in \Omega$$

implies

$$\gamma \in h\Omega(g \exp t_0 X)^{-1} \subset \bigcup_{|t|<\delta} h\Omega(g \exp tX)^{-1} .$$

As the last set is relatively compact, its intersection with Γ is a finite set. Since

$$\lim_{t \to 0} (X\alpha)(h^{-1}\gamma g \exp t_0 X) = (X\alpha)(h^{-1}\gamma g)$$

for each $\gamma \in \Gamma$, we have

$$\lim_{t \to 0} \sum_{\gamma \in \Gamma} (X\alpha)(h^{-1}\gamma g \exp t_0 X)\cdot f(h) = \sum_{\gamma \in \Gamma} (X\alpha)(h^{-1}\gamma g)\cdot f(h) .$$

By Lemma 2.3.1, there exist constants C, N > 0 such that

$$\sum_{\gamma \in \Gamma} |(X\alpha)(h^{-1}\gamma g \exp t_0 X)| \leq C \| g \exp t_0 X \|^N$$

$$\leq C \| g \|^N \| \exp t_0 X \|^N ,$$

where $|t_0| \leq |t| \leq \delta$; and since $f \in L^2(\Gamma \setminus G) \subset L^1(\Gamma \setminus G)$, it follows from the dominated convergence theorem that

$$X(\alpha * f)(g) = (X\alpha) * f(g) .$$

Next,

$$X(\alpha * f^{\mathrm{r}}) = \lim_{t \to 0} \int_G (\alpha(h^{-1}g \exp tX) - \alpha(h^{-1}g))/t \;\; f^{\mathrm{r}}(h)\, dh$$

$$= \lim_{t \to 0} \int_G \int_{N_r \cap \Gamma \setminus N_r} (\alpha(h^{-1}g \exp tX) - \alpha(h^{-1}g))/t \;\; f(nh)\, dndh$$

$$= \lim_{t \to 0} \int_{N_r \cap \Gamma \setminus N_r} \int_G (\alpha(h^{-1}g \exp tX) - \alpha(h^{-1}g))/t \;\; f(nh)\, dndh$$

$$= \lim_{t \to 0} \int_{N_r \cap \Gamma \setminus N_r} \int_G (\alpha(h^{-1}ng \exp tX) - \alpha(h^{-1}ng))/t \;\; f(h)\, dndh$$

$$= \lim_{t \to 0} \int_{N_r \cap \Gamma \setminus N_r} \int_G (X\alpha)(h^{-1}ng \exp t_0X)f(h)\, dhdn\,.$$

The computation in the first part of this proof implies that

$$\Big| \int_G (X\alpha)(h^{-1}ng \exp t_0X)f(h)\, dh \Big| \le C\, \|g\|^N \|\exp t_0X\|^N \mathrm{Vol}(\Gamma \setminus G)^{1/2} \|f\|_2 \|n\|^N\,,$$

where $|t_0| \le |t| < \delta$. Since $N_r \cap \Gamma \setminus N_r$ is compact, there exists a constant $\kappa > 0$ such that

$$\Big| \int_G (X\alpha)(h^{-1}ng \exp t_0X)f(h)\, dh \Big| \le \kappa\,,$$

where $|t_0| \le |t| < \delta$. Also, we know already that

$$\lim_{t \to 0} \int_G (X\alpha)(h^{-1}ng \exp t_0X)f(h)\, dh = \int_G (X\alpha)(h^{-1}ng)f(h)\, dh \qquad \text{for all } n\,.$$

Thus, by the dominated convergence theorem again,

$$X(\alpha * f^r)(g) = \int_{N_r \cap \Gamma \setminus N_r} \lim_{t \to 0} \int_G (X\alpha)(h^{-1}ng \exp t_0X)f(h)\, dh$$

$$= \int_{N_r \cap \Gamma \setminus N_r} \int_G (X\alpha)(h^{-1}ng)f(h)\, dh = (X\alpha)f^r(g)\,. \quad \square$$

We can deduce from our proposition, via an easy induction, that for all $D \in \boldsymbol{D}(G)$, $D(\alpha * f)$ and $D(\alpha * f^r)$ exist, and are equal to $(D\alpha) * f$ and $(D\alpha) * f^r$ respectively. Also, they are continuous by Proposition 2.3.3.

2.4. Definition of Compact Operators

Let μ denote the Haar measure of G. In this work, a *fundamental domain* for $\Gamma \setminus G$ is a μ-measurable subset of G consisting of exactly one representative from each equivalence class in $\Gamma \setminus G$. With appropriate normalization the measure $\mu_{\Gamma \setminus G}$ it inherits from being identified with $\Gamma \setminus G$ is the same as the restriction of μ to it. (See Raghunathan [1].) We assume that there is a *Siegel domain* S_0 associated with the fixed minimal cuspidal subgroup P_0 which contains a fundamental domain for $\Gamma \backslash G$, and which we shall denote by F. (For the case of $G = SL(n, \mathbf{R})$, $\Gamma = SL(n, \mathbf{Z})$, see Borel [1, p. 17].)

For the rest of this chapter we drop the subscript "0" referring to the minimal cuspidal subgroup P_0. Recall the Langlands decomposition of P:

$$P = NAM .$$

Then

$$S_0 = \Omega_N A(t_0) K ,$$

where $A(t_0) = \{a \in A : \alpha_r(\log a) \leq t_0 \text{ for each } \alpha_r \in \Sigma^o\}$, Σ^o is the set of simple roots of $(\boldsymbol{p}, \boldsymbol{a})$; and Ω_N is a compact subset of N. The component $A(t_0)$ represents the noncompact part, or the "cuspidal part", of S_0, as each simple root α_r takes on value indefinitely small. Our present assumption on $\Gamma \backslash G$ means that there is only one (m-dimensional) "cusp."

The definition of our compact operators is motivated by the estimates of Godement (Godement [2], [3]). For g = nak, let H(g) denote log a. Set

$$\eta(u) = \inf_{1 \leq r \leq m} \alpha_r(u) \qquad (u \in \boldsymbol{a}) .$$

A function f defined on G is said to be *slowly increasing* if there exist $C, C' > 0$ such that

$$|f(g)| \leq C' e^{-C\eta(H(g))} \qquad \text{for all } g \in S_0 .$$

For automorphic functions, the condition of slowly increasing is equivalent to polynomial growth. This will be precisely stated in Lemma 3.3.2.

Lemma 2.4.1. Let α be a continuous compactly supported function on G.

(i) For $f \in L^2(\Gamma \backslash G)$, there exists $M > 0$, which does not depend on f, such that for every $N > 0$ and $r \in \{1, \ldots, m\}$, independent of f,

$$\alpha * f(g) - \alpha * f^r(g) \ll e^{-M\eta(H(g))} e^{N\alpha_r(H(g))} \|f\|_2 \qquad \text{for } g \in S_0 .$$

(ii) Let f be a Γ-automorphic continuous slowly increasing function on G, that is, there exist C, C′ > 0 such that

$$|f(g)| \le C' e^{-C\eta(H(g))} \qquad \text{for } g \in S_0 ,$$

then, depending on C and C′, there exists M > 0 such that for every N > 0 and f ∈ {1, . . . , m},

$$\alpha * f(g) - \alpha * f^{r}(g) \le B e^{-M\eta(H(g))} e^{N\alpha_r(H(g))} \qquad \text{for } g \in S_0 ,$$

where B is a positive constant; the majoration is otherwise independent of f.

We first introduce certain *truncation factors* that will appear in the definition of the compact operator. For $\Sigma^o = \{\alpha_p : p = 1, \ldots, m\}$, where m = dim $\boldsymbol{a}$, let

$$a_p = \{ u \in \boldsymbol{a} : \inf_{1 \le r \le m} \alpha_r(u) = \alpha_p(u) \}, \quad p = 1, \ldots, m .$$

It is easily seen that

$$a_p = \bigcap_{1 \le r \le m} \{ u \in \boldsymbol{a} : \alpha_r(u) - \alpha_p(u) \ge 0 \} ,$$

the sets a_p are therefore closed sets. For every $u \in \boldsymbol{a}$,

$$\inf_{1 \le r \le m} \alpha_r(u) = \alpha_p(u) \qquad \text{for some } p ,$$

hence $\boldsymbol{a} = \bigcup_{P} a_p$. Let

$$c_{i,j} = \{ u \in \boldsymbol{a} : |\alpha_i(u) - \alpha_j(u)| < 1 \}, \quad 1 \le i \ne j < m .$$

It is clear that $a_r \cap a_p \subset c_{p,r}$, $1 \le p \ne r \le m$. All the sets $c_{p,r}$ are open, so for each r, $\bigcup_{p \ne r} (a_p \setminus c_{p,r})$ is closed. Hence a_r and $\bigcup_{p \ne r} (a_p \setminus c_{p,r})$ are disjoint closed sets for each r.

We now let $\{h_r(u)\}$ be a set of continuous functions on $\boldsymbol{a}$ such that $0 \le h_r(u) \le 1$ and

$$h_r(u) = \begin{cases} 1 & \text{on } \mathfrak{a}_r \\ 0 & \text{on } \bigcup_{p \neq r} (\mathfrak{a}_p \setminus c_{p,r}) . \end{cases}$$

Note that since $\boldsymbol{a} = \bigcup_p \mathfrak{a}_p$, $h_1(u) + \cdots + h_m(u) \neq 0$ for all $u \in \boldsymbol{a}$. Define

$$H_r(u) = \frac{h_r(u)}{h_1(u) + \cdots + h_m(u)} \qquad (u \in \boldsymbol{a}), \quad r = 1, \ldots, m .$$

If $u \in \mathfrak{a}_r \setminus \bigcup_{j \neq r} c_{r,j}$, by construction $h_p(u) = 0$ for $p \neq r$, then $H_r(u) = 1$. If $u \in \bigcup_{p \neq r} (\mathfrak{a}_p \setminus c_{p,r})$, by construction $h_r(u) = 0$, then $H_r(u) = 0$. Thus the continuous functions $H_r(u)$ on $\boldsymbol{a}$ have the properties that $0 \leq H_r(u) \leq 1$,

$$H_r(u) = \begin{cases} 1 & \text{on } \mathfrak{a}_r \setminus \bigcup_{j \neq r} c_{r,j} \\ 0 & \text{on } \bigcup_{p \neq r} (\mathfrak{a}_p \setminus c_{p,r}) , \end{cases}$$

and $H_1(u) + \cdots + H_m(u) = 1$ for all $u \in \boldsymbol{a}$. By virtue of the logarithmic mapping $H_r(u)$ correspond to continuous functions $\kappa_r(a)$ on A. Define

$$\delta_r(g) = \kappa_r(a) \qquad \text{for } g = nak .$$

The truncation factors δ_r are continuous functions on G, and

$$\sum_{r=1}^{m} \delta_r(g) = 1 \qquad \text{for all } g \in G .$$

Next we derive a consequence of Lemma 2.4.1.

<u>Lemma 2.4.2.</u> Let α be a continuous compactly supported function on G.

(i) There exists $\kappa > 0$ such that for all $f \in L^2(\Gamma \backslash G)$,

$$|\alpha * f(g) - \sum_{r=1}^{m} \delta_r(g)\alpha * f^r(g)| \leq \kappa \| f \|_2 \qquad \text{for } g \in S_0 .$$

(ii) Let f be a Γ-automorphic continuous slowly increasing function on G, that is, there exist $C, C' > 0$ such that

$$|f(g)| \leq C' e^{-C\eta(H(g))} \qquad \text{for } g \in S_0 .$$

Then, depending on C and C', there exists $\kappa > 0$ such that

$$|\alpha * f(g) - \sum_{r=1}^{m} \delta_r(g)\alpha * f^r(g)| \leq \kappa \qquad \text{for } g \in S_0 ,$$

where κ is otherwise independent of f.

<u>Proof</u>. We give only the proof of (i); the proof of (ii) is entirely analogous. Since $\sum_{r=1}^{m} \delta_r(g) = 1$,

$$|\alpha * f(g) - \sum_{r=1}^{m} \delta_r(g)\alpha * f^r(g)|$$

$$= |\sum_{r=1}^{m} \delta_r(g)(\alpha * f(g) - \alpha * f^r(g))| .$$

Before applying Lemma 2.4.1(i), we introduce some notations:

$$A^{(p)} = \exp(a_p), \qquad A^{(i,j)} = \exp(c_{i,j}),$$

$$S_0^{(p)} = \Omega_N(A(t_0) \cap A^{(p)})K , \quad S_0^{(i,j)} = \Omega_N(A(t_0) \cap A^{(i,j)})K .$$

By virtue of their definition the functions $\delta_r(g)$ have the property that $0 \leq \delta_r(g) \leq 1$ and

$$\delta_r(g) = \begin{cases} 1 & \text{on } S_0^{(r)} \setminus \bigcup_{j \neq r} S_0^{(r,j)} \\ 0 & \text{on } \bigcup_{p \neq r} (S_0^{(p)} \setminus S_0^{(p,r)}) . \end{cases}$$

Fix $r \in \{1, \ldots, m\}$ and consider $|\delta_r(g)(\alpha * f(g) - \alpha * f^r(g))|$ for $g \in S_0$. It is clear that $S_0 = \bigcup_p S_0^{(p)}$. Either $g \in S_0^{(r)}$, or $g \in S_0^{(p)}$ for some $p \neq r$.

In Lemma 2.4.1(i), choose $N > 0$ such that $N - M > 0$. Suppose $g \in S_0^{(r)}$. Then $\eta(H(g)) = \alpha_r(H(g))$. Therefore

$$\begin{aligned} e^{-M\eta(H(g))} e^{N\alpha_r(H(g))} &= e^{(N-M)\alpha_r(H(g))} \\ &\leq e^{(N-M)t_0} \end{aligned}$$

because $g \in S_0$. By Lemma 2.4.1(i), we have

$$|\delta_r(g)(\alpha * f(g) - \alpha * f^r(g))| \leq Be^{(N-M)t_0} \|f\|_2 \qquad \text{for some } B > 0 .$$

On the other hand, suppose instead $g \in S_0^{(p)}$ for some $p \neq r$. Then $\eta(H(g)) = \alpha_p(H(g))$. Either $g \in S_0^{(p)} \setminus S_0^{(p,r)}$ or $g \in S_0^{(p,r)}$. In the first case, $\delta_r(g) = 0$, so $|\delta_r(g)(\alpha * f(g) - \alpha * f^r(g))| = 0$. Whereas if $g \in S_0^{(p,r)}$, $\alpha_r(H(g)) - \alpha_p(H(g)) < 1$, Lemma 2.4.1(i) then yields

$$\begin{aligned} &|\delta_r(g)(\alpha * f(g) - \alpha * f^r(g))| \\ &\qquad \leq |\alpha * f(g) - \alpha * f^r(g)| \leq Ce^{-M\alpha_p(H(g)) + N\alpha_r(H(g))} \|f\|_2 \\ &\qquad \leq Ce^{-M\alpha_p(H(g)) + N(1+\alpha_p(H(g)))} \|f\|_2 = Ce^{(N-M)\alpha_p(H(g)) + N} \|f\|_2 \\ &\qquad \leq Ce^{(N-M)t_0+N} \|f\|_2 \qquad \text{for some constant } C . \end{aligned}$$

Therefore, for some $D > 0$,

$$|\delta_r(g)(\alpha * f(g) - \alpha * f^r(g))| \leq D \|f\|_2 \text{ for } g \in S_0 .$$

It is now clear that there exists $\kappa > 0$ such that

$$|\sum_{r=1}^{m} \delta_r(g)(\alpha * f(g) - \alpha * f^r(g))| \leq \kappa \|f\|_2 \quad \text{for } g \in S_0. \quad \square$$

Before presenting the definition of the compact operators on $L^2(\Gamma \setminus G)$, let us recall a standard definition in the spectral theory of compact operators on Banach spaces. Let X and Y be *Banach spaces*. A bounded operator $T\colon X \to Y$ is said to be *compact* if every bounded sequence $\{x_i\}$ in X contains a subsequence $\{x_{i_k}\}$ such that $\{Tx_{i_k}\}$ converges in Y. (See Rudin [1, pp. 97-98]; Yosida [1, p. 277].)

If h is a function defined on F, the fixed fundamental domain for $\Gamma \setminus G$, let $[\,h\,]_F$ denote the Γ-*automorphic extension* of h from F to all of G; if h is defined on a set containing F, we shall write $[\,h\,]_F$ for $[\,h|_F\,]_F$.

The compact operators for the analytic continuation of the Eisenstein series are of the following type. Let α be a smooth compactly supported function on G. Define at operator $\mathbf{K}_\alpha$ on $L^2(\Gamma \setminus G)$ as follows: for $f \in L^2(\Gamma \setminus G)$, let

$$\mathbf{K}_\alpha^o f = \left[\alpha * f - \sum_{r=1}^{m} \delta_r \alpha * f^r\right]_F$$

and

$$\mathbf{K}_\alpha = \lambda_\alpha \mathbf{K}_\alpha^o,$$

where $\lambda_\alpha = \alpha * \alpha *$, the 2-fold convolution with α. (For simplicity, we have written $\delta_r \alpha * f^r$ for $\delta_r(\alpha * f^r)$. Also, we shall sometimes drop the subscript "α" in $\mathbf{K}_\alpha^o$, $\mathbf{K}_\alpha$.)

The main result of this section is Theorem 2.4.1.

Theorem 2.4.1. The mapping $\mathbf{K}_\alpha$ is a compact operator from $L^2(\Gamma \setminus G)$ onto $L^2(\Gamma \setminus G)$.

Before presenting the proof of this theorem, some preliminary considerations about the definition of $\mathbf{K}_\alpha$ are in order.

Proposition 2.4.1. For $f \in L^2(\Gamma \backslash G)$, $\mathbf{K}_\alpha^o f$ is a Γ-automorphic bounded measurable function on G, and hence belongs to $L^2(\Gamma \backslash G)$.

Proof. By its definition, $\mathbf{K}_\alpha^o f$ is Γ-automorphic. Since $F \subset S_0$, by Lemma 2.4.2(i), there exists $\kappa > 0$ such that

$$|\mathbf{K}_\alpha^o f(g)| \leq \kappa \| f \|_2 \qquad \text{for all } g \in G .$$

(In fact, this holds for all $f \in L^2(\Gamma \backslash G)$.) Clearly

$$\mathbf{K}_\alpha^o f = \left[\alpha * f - \sum_{r=1}^{m} \delta_r \alpha * f^r \right]_F = \alpha * f - \sum_{r=1}^{m} [\, \delta_r \alpha * f^r \,]_F \ .$$

As before μ denotes the Haar measure on G. It follows from Proposition 2.3.3 that $\alpha * f$, $\delta_r \alpha * f^r$ $(r = 1, \ldots, m)$ are all μ-measurable. We fix $r \in \{1, \ldots, m\}$ and verify that $[\, \delta_r \alpha * f^r \,]_F$ is measurable. For simplicity, let T denote $\delta_r \alpha * f^r$.

Let U be an open subset of $\mathbf{C}$, we need to check that $[\,T\,]_F^{-1}(U)$ is μ-measurable. First of all F is μ-measurable. Now

$$[\,T\,]_F^{-1}(U) = \bigcup_{\gamma \in \Gamma} \gamma \,(T^{-1}(U) \cap F \,) .$$

It is clear that $T^{-1}(U) \cap F$ is measurable, and $\gamma\,(T^{-1}(U) \cap F\,)$ is measurable for all $\gamma \in \Gamma$. Since Γ is countable, $[\,T\,]_F^{-1}(U)$ too is measurable. So $\mathbf{K}_\alpha^o f$ is a Γ-automorphic measurable function on G. Besides being μ-measurable, when identified with $\Gamma \backslash G$, the fundamental domain F inherits the measure $\mu_{\Gamma \backslash G}$ of $\Gamma \backslash G$, which is the same as the restriction of μ to F. It follows from this that on $\Gamma \backslash G$ the function $\mathbf{K}_\alpha^o f$ is $\mu_{\Gamma \backslash G}$-measurable. Therefore, it belongs to $L^2(\Gamma \backslash G)$.

□

Recall Proposition 2.3.2, and we deduce from the preceding proof the next proposition.

Proposition 2.4.2. There exists $\kappa > 0$ such that for all $f \in L^2(\Gamma \backslash G)$,

$$|\mathbf{K}_\alpha f(g)| \leq \kappa \|f\|_2 \qquad \text{for all } g \in G.$$

Note that $\mathbf{K}_\alpha f$ is Γ-automorphic by Proposition 2.3.1, measurable by virtue of Proposition 2.3.3. So indeed $\mathbf{K}_\alpha : L^2(\Gamma \backslash G) \to L^2(\Gamma \backslash G)$.

We now turn to consider Theorem 2.4.1. Lemma 2.4.3 is the crucial step.

Lemma 2.4.3. Let $\{f_i\}$ be a bounded sequence in $L^2(\Gamma \backslash G)$. For every $g \in G$, the set of functions $\{\mathbf{K}_\alpha f_i\}$ is equicontinuous at g, i.e., for all $\varepsilon > 0$, there exists a neighborhood V of g such that $z \in V$ implies

$$|\mathbf{K}_\alpha f_i(z) - \mathbf{K}_\alpha f_i(g)| < \varepsilon \qquad \text{for all } i.$$

Proof. It suffices to prove equicontinuity at every $g \in F$. Let $\{X_1, \ldots, X_n\}$ be a basis for $\mathfrak{g}$, the Lie algebra of G. Let $b > 0$ be small enough so that

$$N = \{ X \in \mathfrak{g} : X = \beta_1 X_1 + \cdots + \beta_n X_n, \ |\beta_l| < b \ (1 \leq l \leq n) \}$$

corresponds to a relatively compact neighborhood W of g under the exponential mapping $g \exp : X \mapsto g \exp X$, which acts as a homeomorphism on N. By definition,

$$\mathbf{K}_\alpha f_i = \lambda_\alpha \mathbf{K}_\alpha^o f_i = \alpha * (\alpha * \mathbf{K}_\alpha^o f_i),$$

where $\alpha * \mathbf{K}_\alpha^o f_i$ is in $L^2(\Gamma \backslash G)$. In view of Proposition 2.3.4 and the remark that followed, $\mathbf{K}_\alpha f_i(g \exp tX)$ as a function on the interval $(-1, 1)$ is of class C^1, and the mean value theorem of calculus give

$$\mathbf{K}_\alpha f_i(g \exp tX) = \mathbf{K}_\alpha f_i(g) + [\frac{d}{dt'} \mathbf{K} f_i(g \exp t'X)]_{t'=t_0} t$$

$$= \mathbf{K}_\alpha f_i (g) + [\, X(\mathbf{K}f_i) (g \exp t_0 X) \,] t ,$$

where $|t_0| \le |t| \le 1$. Therefore,

$$|\mathbf{K}_\alpha f_i (g \exp tX) - K_\alpha f_i (g)| = |X(\mathbf{K}_\alpha f_i) (g \exp t_0 X)|\, |t| \,.$$

For $X = \beta_1 X_1 + \cdots + \beta_n X_n$,

$$X(\mathbf{K}_\alpha f_i) = \sum_{l=1}^{n} \beta_l X_l (\mathbf{K}_\alpha f_i) \,.$$

By Proposition 2.3.4,

$$X_l (\mathbf{K}_\alpha f_i) = X_l \alpha * (\alpha * \mathbf{K}_\alpha^{o} f_i) \,.$$

Of course, for any given l, $X_l \alpha$ too is a smooth compactly supported function on G. Once again, by lemma 2.4.2(i) and Proposition 2.3.2, there exists $\kappa_l > 0$ such that for all i,

$$|X_l (\mathbf{K}_\alpha f_i) (g)| \le \kappa_l \, \|f_i\|_2 \qquad \text{for all } g \in G \,.$$

For $X \in N$, $|\beta_l| < b$ $(1 \le l \le n)$, and by hypothesis $\{\|f_i\|_2\}$ is bounded. Hence $X(\mathbf{K}_\alpha f_i)$ are uniformly bounded for $X \in N$ and for all i. Therefore, there exists $k > 0$ such that for all i,

$$|\mathbf{K}_\alpha f_i (g \exp tX) - \mathbf{K}_\alpha f_i (g)| \le k|t| \quad \text{for all } X \in N, \ |t| < 1 \,.$$

Finally, we let V be the image of $(\varepsilon/\varepsilon{+}k)N$ under the homeomorphism g exp. If $z \in V$, $z = g \exp((\varepsilon/\varepsilon{+}k)N)$ for $X \in N$. Thus

$$|\mathbf{K}_\alpha f_i (z) - \mathbf{K}_\alpha f_i (g)| = |\mathbf{K}_\alpha f_i (g \exp \frac{\varepsilon}{\varepsilon+k} X) - \mathbf{K}_\alpha f_i (g)|$$

$$\le k \frac{\varepsilon}{\varepsilon+k} < \varepsilon \,. \quad \square$$

The proof of Theorem 2.4.1 is as follows.

Proof of Theorem 2.4.1. It is an immediate consequence of Proposition 2.4.2 that **K** is a bounded operator on $L^2(\Gamma \backslash G)$. Let $\{f_i\}$ be a L^2-bounded sequence in $L^2(\Gamma \backslash G)$. We want to prove that $\{f_i\}$ contains a subsequence $\{f_{i_k}\}$ such that $\{\mathbf{K}f_{i_k}\}$ converges in $L^2(\Gamma \backslash G)$. Take a countable dense subset $E \subset G$, $E = \{g_1, g_2, \cdots\}$. Because $\{\|f_i\|_2\}$ is bounded, we have from Proposition 2.4.2 that $\{\mathbf{K}f_i(g_j)\}$ is bounded for each $g_j \in E$. By the Ascoli selection we have a subsequence $\{f_{i_k}\}$ of $\{f_i\}$ such that $\{\mathbf{K}f_{i_k}(g_j)\}$ converges for every $g_j \in E$. Now let $g \in G$ be arbitrary. We have

$$|\mathbf{K}f_{i_{k_1}}(g) - \mathbf{K}f_{i_{k_2}}(g)| \le |\mathbf{K}f_{i_{k_1}}(g) - \mathbf{K}f_{i_{k_1}}(g_j)| + |\mathbf{K}f_{i_{k_1}}(g_j) - \mathbf{K}f_{i_{k_2}}(g_j)| + |\mathbf{K}f_{i_{k_2}}(g_j) - \mathbf{K}f_{i_{k_2}}(g)| .$$

By Lemma 2.4.3, for any $\varepsilon > 0$, there exists a neighborhood V of g such that for all i_k,

$$|\mathbf{K}f_{i_k}(g) - \mathbf{K}f_{i_k}(z)| < \frac{\varepsilon}{3} \quad \text{if } z \in V .$$

It follows from the definition of E that V contains a point g_j of E. So

$$|\mathbf{K}f_{i_{k_1}}(g) - \mathbf{K}f_{i_{k_2}}(g)| < \frac{\varepsilon}{3} + |\mathbf{K}f_{i_{k_1}}(g_j) - \mathbf{K}f_{i_{k_2}}(g_j)| + \frac{\varepsilon}{3} .$$

Since $\{\mathbf{K}f_{i_k}(g_j)\}$ converges, for k_1 and k_2 both sufficiently large, we have from the above that

$$|\mathbf{K}f_{i_{k_1}}(g) - \mathbf{K}f_{i_{k_2}}(g)| < \varepsilon .$$

Thus $\{\mathbf{K}f_{i_k}(g)\}$ is a Cauchy sequence. Define

$$\Lambda(g) = \lim_{k \to \infty} \mathbf{K}f_{i_k}(g) \quad \text{for } g \in G .$$

Since $\mathbf{K}f_{i_k}$ are Γ-automorphic, so is Λ. It remains to prove that $\Lambda \in L^2(\Gamma \backslash G)$ and

$$\lim_{k \to \infty} \|\mathbf{K}f_{i_k} - \Lambda\|_2 = 0 .$$

Using Proposition 2.4.2 once more, the boundedness of $\{\|f_i\|_2\}$ implies the uniform boundedness of $\mathbf{K}f_{i_k}$, and thence the boundedness of Λ. Then $\mathrm{Vol}(\Gamma \backslash G) < \infty$ implies $\Lambda \in L^2(\Gamma \backslash G)$. Lastly,

$$\lim_{k \to \infty} |\mathbf{K}f_{i_k}(g) - \Lambda(g)|^2 = 0 \quad \text{for all } g \in G$$

and

$$|\mathbf{K}f_{i_k}(g) - \Lambda(g)|^2 \le (|\mathbf{K}f_{i_k}(g)| + |\Lambda(g)|)^2,$$

which is bounded. The dominated convergence theorem then asserts that

$$\lim_{k\to\infty} \int_{\Gamma\backslash G} |\mathbf{K}f_{i_k}(g) - \Lambda(g)|^2 \, dg = 0,$$

i.e.,

$$\lim_{k\to\infty} \|\mathbf{K}f_{i_k} - \Lambda\|_2 = 0. \qquad \square$$

CHAPTER 3

FREDHOLM EQUATIONS

3.1. Introduction

The application of the compact operator $\mathbf{K}_\alpha$ to the Eisenstein series yields

$$(\mathbf{K}_\alpha - \hat{\alpha}(\Lambda))E(\Lambda, g) = -\sum_J \phi(J \mid \Lambda)\hat{\alpha}^o(\Lambda)H_J^o(\Lambda, g)$$

where $\hat{\alpha} = (\hat{\alpha}^o)^3$, and $\phi(J \mid \Lambda)$, $H_J^o(\Lambda, g)$ are functions derived from the constant terms of the Eisenstein series. The constant term coefficients $\phi(J \mid \Lambda)$ are defined and holomorphic on $(\boldsymbol{a}_C^*)^+$, where the Eisenstein series was originally defined (Section 3.2). Each $H_J^o(\Lambda, g) \in C^\infty(\boldsymbol{a}_C^* \times G)$ and is holomorphic on $\boldsymbol{a}_C^*$ for all $g \in G$ (Theorem 3.4.1).

The basic Fredholm equations are:

$$(\mathbf{K}_\alpha - \hat{\alpha}(\Lambda))^oF_J^{**}(\Lambda) = -\mathbf{K}_\alpha H_J^o(\Lambda) \qquad (\Lambda \in \boldsymbol{a}_C^* \setminus \hat{\alpha}^{-1}(\text{spec } \mathbf{K}_\alpha)) .$$

With the use of reduction theory, we shall show that $H_J^o(\Lambda, g)$ is slowly increasing. Then it follows from Godement's second estimate (Lemma 2.4.2(ii)) that $\mathbf{K}_\alpha H_J^o(\Lambda, g) \in L^2(\Gamma \setminus G)$ (Theorem 3.4.2). Truncation of the Eisenstein series will proceed along the same line (Theorem 3.5.1). In addition, $\mathbf{K}_\alpha H_J^o(\Lambda, g) \in C^\infty(\boldsymbol{a}_C^* \times G)$ and is holomorphic on $\boldsymbol{a}_C^*$ for each $g \in G$ (Theorem 3.4.2). The smoothness and holomorphy of $\mathbf{K}_\alpha H_J^o(\Lambda, g)$ ensure that ${}^oF_J^{**}(\Lambda, g)$ have similar properties (Corollary 1 to Theorem 3.6.1). These are, of course, indispensable in establishing the analytic continuation of the Eisenstein series. We shall discuss this in the next chapter.

It should be pointed out that the standard theory of compact operators will only give us L^2-classes as Fredholm solutions. That nice pointwise defined solutions do exist requires some verification (Section 3.6). The key facts are, once again, Godement's estimates.

3.2. The Constant Terms of Eisenstein Series

The compact operator $\mathbf{K}_\alpha$ defined in the last chapter involves every constant term with respect to a maximal standard cuspidal subgroup. In this section we present Langlands' formula for the constant terms of a cuspidal Eisenstein series. Knowledge of the form of the constant terms is crucial to our development, as $\mathbf{K}_\alpha$ will eventually be applied to the Eisenstein series.

Recall that $E(\Lambda, \Phi, g)$ is a cuspidal Eisenstein series for the maximal standard cuspidal subgroup P. Proofs of Lemma 3.2.1 and Theorem 3.2.1 can be found in Harish-Chandra [2], Chapter 2, Sections 4-5.

Lemma 3.2.1. Suppose P_r is a maximal standard cuspidal subgroup not associate to P, then $E^r(\Lambda, g) = 0 \ \ ((\Lambda, g) \in (\boldsymbol{a}_C^*)^+ \times G)$.

From now on $\{P_j\}$ denotes the subset of $\{P_r\}$ consisting of all maximal standard cuspidal subgroups associate to P. Next we present Langlands' formula.

Theorem 3.2.1. Suppose P_j is a maximal standard cuspidal subgroup associate to P, then

$$E^j(\Lambda, g) = \sum_{s \in W(\boldsymbol{a}, \boldsymbol{a}_j)} e^{(-s\Lambda - \rho_j)(\log a_j)} \sigma(k) M(s \mid \Lambda) \Phi(m_j)$$

$(\Lambda \in (a_C^*)^+ \,;\, g = n_j a_j m_j k,\ n_j \in N_j,\ a_j \in A_j,\ m_j \in M_j,\ k \in K)$, where $W(\boldsymbol{a}, \boldsymbol{a}_j)$ is the *Weyl group* of $(\boldsymbol{a}, \boldsymbol{a}_j)$; $M(s \mid \Lambda)$ is a linear transformation from ${}^oL^2(\Gamma_M \backslash M, \sigma_M, \chi)$ into ${}^oL^2(\Gamma_{M_j} \backslash M_j, \sigma_{M_j}, {}^s\chi)$ (cf. Section 1.2 for definition of ${}^oL^2(\Gamma_M \backslash M, \sigma_M, \chi)$ and ${}^oL^2(\Gamma_{M_j} \backslash M_j, \sigma_{M_j}, {}^s\chi)$).

In the last statement ${}^s\chi$ is the character of $\mathbf{Z}(M_j)$ defined by

$${}^s\chi(D_{M_j}) = \chi({}^{s^{-1}}D_{M_j}) \qquad (D_{M_j} \in \mathbf{Z}(M_j)) \,.$$

Each $s \in W(\boldsymbol{a}, \boldsymbol{a}_j)$ induces an isomorphism from $\mathbf{Z}(M)$ onto $\mathbf{Z}(M_j) : D_M \mapsto {}^sD_M$. Note that, though the notation is not explicit, the linear transformation $M(s\,|\,\Lambda)$ depends on χ, a specific character of $\mathbf{Z}(M)$.

The *intertwining operators* $M(s\,|\,\Lambda)$ are holomorphic in Λ for $\Lambda \in (\mathfrak{a}_{\mathbb{C}}^*)^+$. It is known that ${}^oL^2(\Gamma_M \backslash M, \sigma_M, \chi)$ and ${}^oL^2(\Gamma_{M_j} \backslash M_j, \sigma_{M_j}, {}^s\chi)$ are finite dimensional vector spaces (Harish-Chandra [2], Chapter 1, Section 6). If $\{\Phi_p\}_p$ and $\{\Psi_{j,s,n}\}_n$ are respectively bases of ${}^oL^2(\Gamma_M \backslash M, \sigma_M, \chi)$ and ${}^oL^2(\Gamma_{M_j} \backslash M_j, \sigma_{M_j}, {}^s\chi)$, and $M(s\,|\,\Lambda)$ has the matrix representation $[\theta(j, s, n; p\,|\,\Lambda)]$, then for

$$\Phi = \sum_p \eta_p \Phi_p \,,$$

we have

$$M(s\,|\,\Lambda)\Phi = \sum_n \phi(j, s, n\,|\,\Lambda)\Psi_{j,s,n} \,,$$

where $\phi(j, s, n\,|\,\Lambda) = \sum_p \eta_p \theta(j, s, n; p\,|\,\Lambda)$. The constant term coefficients $\phi(j, s, n\,|\,\Lambda)$ are defined and holomorphic on $(\mathfrak{a}_{\mathbb{C}}^*)^+$. In terms of these, Langland's formula reads:

$$E^j(\Lambda, g) = \sum_{s,n} \phi(j, s, n\,|\,\Lambda) e^{(-s\Lambda-\rho_j)(\log a_j)} \sigma(k) \Psi_{j,s,n}(m_j) \,.$$

Note that insofar as the constant terms are concerned, the difficulty of analytic continuation lies in the coefficients.

3.3. Convolution with Functions Depending on a Complex Parameter

Though first defined on $L^2(\Gamma \backslash G)$, the compact operator $\mathbf{K}_\alpha$ of the last chapter will be applied to the Eisenstein series, and from this a set of Fredholm equations will result. This requires the convolution of α with Γ-automorphic functions on G which also depend holomorphically on a complex parameter. This section is devoted to the presentation of some technical results in this direction. Most of these point to the smoothing effect of convolution.

Proposition 3.3.1. Let α be a continuous compactly supported function on G. Let U be an open subset of C, and suppose for each $z \in U$, f(z, g) is a Γ-automorphic measurable function on G, and for each $g \in G$ a holomorphic function on U. If f(z, g) is bounded on every compact subset of $U \times G$, then $\alpha * f(z, g)$ and $(d^v\alpha * f/dz^v)(z, g)$ are Γ-automorphic functions in $C(U \times G)$, and are holomorphic on U for each $g \in G$.

Proof. By definition,

$$\alpha * f(z, g) = \int_G \alpha(h^{-1}g) f(z, h)\, dh .$$

It follows easily from our hypothesis that $\alpha * f(z, g)$ is well defined and Γ-automorphic. Fix $z_0 \in U$. There exists a neighborhood $\tilde{U}$ of z_0 in U such that

$$f(z, g) = \sum_{k=0}^{\infty} a_k(g)(z - z_0)^k \qquad ((z, g) \in \tilde{U} \times G),$$

where

$$a_k(g) = (2\pi i)^{-1} \int_\beta u^{-k-1} f(u, g)\, du \qquad (g \in G),$$

β being a counterclockwise circle in $\tilde{U}$ of radius r centered at z_0. The functions $a_k(g)$ are measurable. Fix $g_0 \in G$ and let W be a relatively compact neighborhood of g_0. If Ω denotes the compact support of α, then $h^{-1}g \neq 0$ implies $h \in g\Omega^{-1}$. Thus, for $g \in W$, $\alpha(h^{-1}g) \neq 0$ implies $h \in \overline{W\Omega^{-1}}$. Denote the range of β by β^*. Now since f(z, h) is bounded on $\beta^* \times \overline{W\Omega^{-1}}$, there exists $M > 0$ such that

$$| a_k(h) | \leq \frac{M}{r^k} \qquad (h \in \overline{W\Omega^{-1}}).$$

Hence there is a neighborhood V of z_0 in $\tilde{U}$ such that

$$f(z, h) = \sum_{k=0}^{\infty} (z - z_0)^k a_k(h) \qquad \text{uniformly on } V \times \overline{W\Omega^{-1}} .$$

We can write

$$\alpha * f(z, g) = \int_{\overline{W}\Omega^{-1}} \alpha(h^{-1}g)f(z, h)\, dh \qquad ((z, g) \in U \times W),$$

and then

$$\alpha * f(z, g) = \sum_{k=0}^{\infty} (z - z_0)^k \left[\int_{\overline{W}\Omega^{-1}} \alpha(h^{-1}g)a_k(h)\, dh \right] \qquad ((z, g) \in V \times W).$$

For $g \in W$,

$$\left| \int_{\overline{W}\Omega^{-1}} \alpha(h^{-1}g)a_k(h)\, dh \right| \leq \frac{M}{r^k} \int_{\overline{W}\Omega^{-1}} |\alpha(h^{-1}g)|\, dh = \frac{M}{r^k} \int_G |\alpha(h^{-1}g)|\, dh .$$

Under the change of variable $g^{-1}h \mapsto h$, the last quantity becomes

$$\frac{M}{r^k} \int_G |\alpha(h^{-1})|\, dh \leq \frac{M}{r^k} \int_{\Omega^{-1}} |\alpha(h^{-1})|\, dh \leq \frac{M'}{r^k} \qquad \text{for some } M' > 0 .$$

So it may be supposed that the infinite series for $\alpha * f(z, g)$ too converges uniformly on $V \times W$. It is easily verified that for each k,

$$\int_{\overline{W}\Omega^{-1}} \alpha(h^{-1}g)a_k(h)\, dh$$

defines a continuous function on W. Therefore, $\alpha * f(z, g)$ is a continuous function on $V \times W$, and is holomorphic on V for each $g \in W$. For $(d^\nu \alpha * f/dz^\nu)(z, g)$ we perform repeated termwise differentiation on the power series for $\alpha * f(z, g)$. The resulting series will also be uniformly convergent on $V \times W$. So $(d^\nu \alpha * f/dz^\nu)(z, g)$ are continuous on $V \times W$, and are holomorphic in V for each $g \in W$ also. Since $z_0 \in U$ and $g_0 \in G$ are arbitrary, our proof is complete. □

Proposition 3.3.2. Let α be a smooth compactly supported function on G. Let U be an open subset of **C**, and suppose for each $z \in U$, f(z, g) is a Γ-automorphic function on G, and for each $g \in G$ a holomorphic function on U. If f(z, g) and $(d^\nu f/dz^\nu)(z, g)$ $(\nu = 1, 2, \cdots)$ all belong to $C(U \times G)$, then $\alpha * f(z, g)$ is in $C^\infty(U \times G)$, and is holomorphic on U for each $g \in G$.

Proof. Let $z = x + iy$. From the hypothesis we know that

$$\frac{\partial^{r+s} f}{\partial x^r \partial y^s}(x, y; g), \qquad (r, s) \in (\mathbf{Z}^+)^2,$$

all exist and are continuous on $U \times G$. We have

$$\alpha * f(z, g) = \int_G \alpha(h^{-1}g) f(z, h)\, dh.$$

Fix $g_0 \in G$ and $z_0 = x_0 + iy_0 \in U$. Fix a coordinate neighborhood of g_0 and let W be a relatively compact neighborhood of g_0 contained in it. Let O be a relatively compact neighborhood of z_0 in U. We prove first that $(\partial^{r+s}\alpha * f/\partial x^r \partial y^s)(x, y; g)$ all exist and are continuous on $O \times W$. For $g \in W$, $\alpha(h^{-1}g) \neq 0$ implies $h \in \overline{W\Omega^{-1}}$, where Ω is the compact support of α. For $(x, y; g) \in O \times W$,

$$\lim_{\Delta x \to 0} (\alpha * f(x + \Delta x, y; g) - \alpha * f(x, y; g))/\Delta x$$

$$= \lim_{\Delta x \to 0} \int_{\overline{W\Omega^{-1}}} \alpha(h^{-1}g)(f(x + \Delta x, y; h) - f(x, y; h))/\Delta x\, dh$$

$$= \lim_{\Delta x \to 0} \int_{\overline{W\Omega^{-1}}} \alpha(h^{-1}g) \frac{\partial f}{\partial x}(x', y; h)\, dh,$$

where $|x - x'| \leq |\Delta x|$. The partial derivative $(\partial f/\partial x)(x, y; h)$ is continuous on $O \times \overline{W\Omega^{-1}}$, so for $(x, y; h) \in O \times \overline{W\Omega^{-1}}$, $\alpha(h^{-1}g)(\partial f/\partial x)(x, y; h)$ is bounded.

An application of the dominated convergence theorems shows that

$$\frac{\partial \alpha * f}{\partial x}(x, y; g) = \int_{\overline{W\Omega^{-1}}} \alpha(h^{-1}g) \frac{\partial f}{\partial x}(x, y; h)\, dh = \int_G \alpha(h^{-1}g) \frac{\partial f}{\partial x}(x, y; h)\, dh$$

$((x, y; g) \in O \times W)$. Therefore $\dfrac{\partial \alpha * f}{\partial x}(x, y; g)$ exists, and is easily seen to be continuous on $O \times W$. Inductively, the same can be established for

$$\frac{\partial^{r+s} \alpha * f}{\partial x^r \partial y^s}(x, y; g), \quad (r, s) \in (\mathbf{Z}^+)^2 \qquad ((x, y; g) \in O \times W).$$

Next, let $V \subset \mathbf{R}^n \to W : (t_1, \ldots, t_n) \mapsto g(t_1, \ldots, t_n)$ be a coordinate mapping $(n = \dim G)$,

$$\lim_{\Delta t_1 \to 0} \{(\alpha * f(z, g(t_1 + \Delta t_1, \ldots, t_n)) - \alpha * f(z, g(t_1, \ldots, t_n))\}/\Delta t_1$$

$$= \lim_{\Delta t \to 0} \int_{\overline{W\Omega^{-1}}} \{\alpha(h^{-1}g(t_1 + \Delta t_1, \ldots, t_n)) - \alpha(h^1 g(t_1, \ldots, t_n))\}/\Delta t_1 \cdot f(z, h)\, dh$$

$$= \lim_{\Delta t \to 0} \int_{\overline{W\Omega^{-1}}} \frac{\partial \alpha}{\partial t_1}(h^{-1}g(t_1', t_2, \ldots, t_n))f(z, h)\, dh,$$

where $|t_1 - t_1'| \le |\Delta t_1|$. Being continuous,

$$\frac{\partial \alpha}{\partial t_1}(h^{-1}g(t_1', t_2, \ldots, t_n))f(z, h)$$

is bounded on $V \times \overline{W\Omega}^{-1}$. It follows from the dominated convergence theorem that

$$\frac{\partial \alpha * f}{\partial t_1}(z, g(t_1, \ldots, t_n)) = \int_{\overline{W\Omega^{-1}}} \frac{\partial \alpha}{\partial t_1}(h^{-1}g(t_1, \ldots, t_n))f(z, h)\, dh$$

$$= \int_G \frac{\partial \alpha}{\partial t_1}(h^{-1}g(t_1, \ldots, t_n))f(z, h)\, dh \qquad ((z, g) \in U \times W).$$

So $\dfrac{\partial \alpha * f}{\partial t_1}(z, g(t_1, \ldots, t_n))$ exists, and is continuous on $U \times W$. Inductively, the same can be established for

$$\frac{\partial^{\lambda_1 + \cdots + \lambda_n} \alpha * f}{\partial_1^{\lambda_1} \cdots \partial t_n^{\lambda_n}}(z, g(t_1, \ldots, t_n)), \qquad (\lambda_1, \ldots, \lambda_n) \in (\mathbf{Z}^+)^n$$

$((z, g(t_1, \ldots, t_n)) \in U \times W)$. Since $(z_0, g_0) \in U \times G$ is arbitrary, the two conclusions have established that $\alpha * f(z, g) \in C^\infty(U \times G)$. Finally, note that α and $f(z, g)$ satisfy the hypothesis of Proposition 3.3.1, so $\alpha * f(z, g)$ is holomorphic on U for each $g \in G$. □

Proposition 3.3.2 will still be valid of $\mathbf{K}_\alpha f$ takes the place of $\alpha * f$. To prove this, we need a result from reduction theory.

Lemma 3.3.1. Let $\pi: G \to \Gamma \backslash G$ be the canonical projection map. For any compact subset C of $\Gamma \backslash G$, $\pi^{-1}(C) \cap S_0$ is compact.

(See Harish-Chandra [2], p. 71.)

Proposition 3.3.3. Let α be a smooth compactly supported function on G. Let U be an open subset of **C**, and suppose for each $z \in U$, $f(z, g)$ is a Γ-automorphic function on G, and for each $g \in G$ a holomorphic function on U. If $f(z, g)$ and $(d^{\nu}f/dz^{\nu})(z, g)$ $(\nu = 1, 2, \cdots)$ all belong to $C(U \times G)$, then $\mathbf{K}_{\alpha}f(z, g)$ is in $C^{\infty}(U \times G)$, and is holomorphic on U for each $g \in G$.

Proof. We need to verify that $\mathbf{K}_{\alpha}f$ is well defined. By definition,

$$\mathbf{K}_{\alpha}f = \lambda_{\alpha}(\mathbf{K}_{\alpha}^{o}f) = \alpha * \alpha * (\alpha * f - \sum_{r=1}^{m} [\delta_r\alpha * f^r]_F).$$

Fix any $r \in \{1, \ldots, m\}$ and consider $[\delta_r\alpha * f^r]_F$. For simplicity let T denote $\delta_r\alpha * f^r$. The function T is continuous on $U \times G$ by virtue of Proposition 3.3.1. We claim that $[T]_F$ is bounded on every compact subset of $U \times G$. Let E and Ω be compact subsets of U and G respectively. Denote by C the compact subset $\pi(\Omega)$ of $\Gamma \backslash G$. Now

$$\{ [T]_F(z, g) : z \in E, g \in \Omega \} = \{ [T]_F(z, g) : z \in E, g \in \pi^{-1}(C) \cap F \}$$

$$= \{ T(z, g) : z \in E, g \in \pi^{-1}(C) \cap F \} \subset \{ T(z, g) : z \in E, g \in \pi^{-1}(C) \cap S_0 \}.$$

By Lemma 3.3.1, $\pi^{-1}(C) \cap S_0$ is compact. Since T is continuous on $U \times G$, the last set of values of $T(z, g)$ is bounded. It follows that $[T]_F(z, g)$ is bounded on $E \times \Omega$. This proves our claim.

Moreover, for each $z \in U$, $[T]_F(z, g)$ is a Γ-automorphic measurable function on G. For each $g \in G$ it is a holomorphic function on U. Hence, by Proposition 3.3.1, $\alpha * [T]_F(z, g)$ is well defined, it and $(d^{\nu}(\alpha * [T]_F)/dz^{\nu})(z, g)$ $(\nu = 1, 2, \cdots)$ all belong to $C(U \times G)$, as well as being holomorphic on U for $g \in G$. The same can, of course, be said of $\alpha * (\alpha * f)$. Hence

$$\alpha * (\mathbf{K}_\alpha^o f) = \alpha * (\alpha * f) - \sum_{r=1}^{m} \alpha * [\delta_r \alpha * f^r]_F ,$$

where each term on the right-hand side satisfies the hypothesis of Proposition 3.3.2. Thus, by Proposition 3.3.2, we have

$$\mathbf{K}_\alpha f = \lambda_\alpha(\alpha * f) - \sum_{r=1}^{m} \lambda_\alpha[\delta_r \alpha * f^r]_F ,$$

as well as the assertions of the present proposition. □

The hypothesis of Proposition 3.3.3 will be satisfied if for each $z \in U$, $f(z, g)$ is a Γ-automorphic function on G, for each $g \in G$ a holomorphic function on U, and it is in $C^\infty(U \times G)$.

The last proposition of this section indicates that convolution preserves growth. We quote another result from reduction theory.

Lemma 3.3.2. An automorphic function f on G is slowly increasing if and only if there exist $C, N > 0$ such that $|f(g)| \le C\|g\|^N$ for all $g \in G$.

[See Harish-Chandra [2, p. 9].)

With this, Proposition 3.3.4 is easy.

Proposition 3.3.4. Let α be a continuous compactly supported function on G. If an automorphic function f is slowly increasing, then $\alpha * f$ too is slowly increasing.

Proof. Let $C, N > 0$ be such that

$$|f(g)| \le C\|g\|^N \qquad \text{for all } g \in G .$$

As usual Ω denotes the compact support of α. Then

$$|\alpha * f(g)| = |\int_G \alpha(h^{-1}g)f(h)\, dh|$$

$$= \left| \int_G \alpha(h^{-1}) f(gh)\, dh \right| \le \int_G |\alpha(h^{-1})| \; |f(gh)|\, dh$$

$$\le \int_G |\alpha(h^{-1})| C \|gh\|^N dh \le C \|g\|^N \int_G |\alpha(h^{-1})| \; \|h\|^N dh$$

$$= C \|g\|^N \int_{\Omega^{-1}} |\alpha(h^{-1})| \; \|h\|^N dh \le C' \|g\|^N \quad \text{for some } C' > 0 . \quad \square$$

3.4. Projection of the Constant Terms of Eisenstein Series

Recall that the constant term of the Eisenstein series for P with respect to another maximal standard cuspidal subgroup P_r is 0 if P and P_r are not associate. On the other hand, if P and P_j are associate to one another, then

$$E^j(\Lambda, g) = \int_{N_j \cap \Gamma \backslash N_j} E(\Lambda, ng)\, dn$$

$$= \sum_{s,n} \phi(j, s, n \,|\, \Lambda) e^{(-s\Lambda - \rho_j)(\log a_j)} \sigma(k) \Psi_{j,s,n}(m_j)$$

($g = n_j a_j m_j k$, $n_j \in N_j$, $a_j \in A_j$, $m_j \in M_j$, $k \in K$). Let

$$I_{j,s,n}(\Lambda, g) = \delta_j(g) e^{(-s\Lambda - \rho_j)(\log a_j)} \sigma(k) \Psi_{j,s,n}(m_j) \qquad (g \in G, \Lambda \in \boldsymbol{a}_C^*) .$$

For every (j, s, n) define a function $H^o_{j,s,n}(\Lambda, g)$ on $\boldsymbol{a}_C^* \times G$ by

$$H^o_{j,s,n} = \lambda_\alpha [\, I_{j,s,n} \,]_F = \alpha * \alpha * [\, I_{j,s,n} \,]_F \qquad \text{for each } \Lambda .$$

The dual space $\boldsymbol{a}_C^*$ can be identified with **C** by picking a nonzero element of $\boldsymbol{a}_C^*$ to form a basis. It is clear that

$$I_{j,s,n}(\Lambda, g) \in C(\, \boldsymbol{a}_C^* \times G \,)$$

and is holomorphic on $\boldsymbol{a}_C^*$ for each $g \in G$. As in the proof of Proposition 3.3.3, $[\, I_{j,s,n} \,]_F (z, g)$ can be shown to be a Γ-automorphic measurable function on G for each $\boldsymbol{a}_C^*$, which is holomorphic on $\boldsymbol{a}_C^*$ for each $g \in G$; also, it is bounded on every compact subset of $\boldsymbol{a}_C^* \times G$.

Therefore, applying Proposition 3.3.1 to $[\,I_{j,s,n}\,]_F$ and then Proposition 3.3.2 to $\alpha * [\,I_{j,s,n}\,]_F$, we have the following theorem.

Theorem 3.4.1. Each function $H^o_{j,s,n}(\Lambda, g) \in C^\infty(\boldsymbol{a}^*_C \times G)$ and is holomorphic on $\boldsymbol{a}^*_C$ for each $g \in G$.

Of course, $H^o_{j,s,n}(\Lambda, g)$ is Γ-automorphic on G for each $\Lambda \in \boldsymbol{a}^*_C$. Both $H^o_{j,s,n}$ and $\mathbf{K}_\alpha H^o_{j,s,n}$ will appear in Fredholm equations.

For $\mathbf{K}_\alpha H^o_{j,s,n}$ it is crucial that we know it is in $L^2(\Gamma \setminus G)$. (This is what we mean by *projection of constant terms.*) Our tool will, of course, be Lemma 2.4.2(ii), so we turn to examine the growth of $H^o_{j,s,n}$.

First we show that as a function on the Siegel domain S_0, $I_{j,s,n}$ is slowly increasing. Then with some results from reduction theory, we shall deduce that $[\,I_{j,s,n}\,]_F$ too is slowly increasing. Hence $H^o_{j,s,n}$ is slowly increasing by virtue of Proposition 3.3.4.

Lemma 3.4.1 below is a basic fact from structure theory which we shall need in the proof of Proposition 3.4.2. Let Σ be the set of roots of $(\boldsymbol{p}_0, \boldsymbol{a}_0)$. For any subset F of Σ^o and the corresponding standard cuspidal subgroup P_F, with $P_F = N_F A_F M_F$, we have $M_F = N(F)A(F)K(F)$ (Warner [1, p. 74]), where $N(F) \subset N_0$, $A(F) \subset A_0$, $K(F) \subset K$. Denote the corresponding Lie algebras by $\boldsymbol{m}(F)$, $\boldsymbol{n}(F)$, $\boldsymbol{a}(F)$, $\boldsymbol{k}(F)$.

Lemma 3.4.1. The restrictions of the elements in $\Sigma \cap (\sum_{\alpha \in F} Z^+\alpha)$ to $\boldsymbol{a}(F)$ form the set of (positive) roots of $\boldsymbol{a}(F)$ in $\boldsymbol{m}(F)$.

(See Warner [1, pp. 66-74, 81-82.]

Proposition 3.4.1. For each $\Lambda \in \boldsymbol{a}_C^*$, every function $I_{j,s,n}$ is slowly increasing. Moreover, for every compact subset E of $\boldsymbol{a}_C^*$ there exist C, C′ > 0 such that

$$|I_{j,s,n}(\Lambda, g)| \leq C'e^{-C\eta(H_0(g))} \quad \text{for } (\Lambda, g) \in E \times S_0 .$$

Proof. Recall that

$$I_{j,s,n}(\Lambda, g) = \delta_j(g)e^{(-s\Lambda-\rho_j)(\log a_j)}\sigma(k)\Psi_{j,s,n}(m_j) .$$

Since $\Psi_{j,s,n}(m_j)$ is a cusp form, it is bounded. So is $\sigma(k)$. We turn our attention to the exponential $e^{(-s\Lambda-\rho_j)(\log a_j)}$. For every $g \in G$,

$$g = n_0a_0k_0 \qquad (n_0 \in N_0, a_0 \in A_0, k_0 \in K)$$

and

$$g = n_ja_jm_jk \qquad (n_j \in N_j, a_j \in A_j, m_j \in M_j, k \in K).$$

If $m_j = n_0'a_0'k_0'$ ($n_0' \in N(F) \subset N_0$, $a_0' \in A(F) \subset A_0$, $k_0' \in K(F) \subset K$), then

$$g = n_ja_jn_0'a_0'k_0'k = (n_ja_jn_0'a_j^{-1})(a_ja_0')(k_0'k) ,$$

where $n_ja_jn_0'a_j^{-1} \in N_0$, $a_ja_0' \in A_0$, $k_0'k \in K$. The equality $a_0 = a_ja_0'$ implies

$$H_0(g) = H_j(g) + H_0(m_j) \qquad \text{for } g = n_ja_jm_jk$$

($n_j \in N_j$, $a_j \in A_j$, $m_j \in M_j$, $k \in K$; $H_j(g) = \log a_j \in \boldsymbol{a}_j$). Recall that

$$\eta(H(g)) = \inf_{1\leq r\leq m} \alpha_r(H_0(g)) ,$$

$$\{\alpha_1, \ldots, \alpha_m\} = \Sigma^o ,$$

$$S_0 = \Omega_{N_0}A(t_0)K ,$$

where $A_0(t_0) = \{a \in A_0 : \alpha_r(\log a) \leq t_0 \text{ for each } \alpha_r \in \Sigma^o\}$. Note that $H_j(a) \in \boldsymbol{a}_j \subset \boldsymbol{a}_0$, $H_0(g)$ and $H_0(m_j) \in \boldsymbol{a}_0$. As

$$\alpha_r(H_0(g)) = \alpha_r(H_j(g)) + \alpha_r(H_0(m_j)) ,$$

we have

$$\alpha_r(H_g(g)) = \alpha_r(H_0(g)) - \alpha_r(H_0(m_j)) .$$

Let $F = \{\alpha_s\}$ be the subset of Σ^o corresponding to which P_j is defined. Since $H_0(m_j) \in \boldsymbol{a}(F)$, $\alpha_r(H_0(m_j)) = \mu(H_0(m_j))$ for some $\mu \in \boldsymbol{a}(F)^*$. So, by Lemma 3.4.1,

$$\alpha_r(H_0(m_j)) = \sum_s C_{rs}\alpha_s(H_0(m_j))$$

for some real constants C_{rs}, where $\alpha_s(H_j(g)) = 0$. Note that

$$\alpha_s(H_j(g)) = 0 \text{ implies } \alpha_s(H_0(m_j)) = \alpha_s(H_0(g)) .$$

Hence,

$$\alpha_r(H_j(g)) = \alpha_r(H_0(g)) - \sum_s C_{rs}\alpha_s(H_0(g)) \qquad \text{for all } r = 1, \ldots, m .$$

Now

$$\left| e^{(-s\Lambda-\rho_j)(\log a_j)} \right| = e^{(\mathrm{Re}(-s\Lambda)-\rho_j)(\log a)} .$$

Let $\{\alpha_t\} \subset \{\alpha_r\}$ be the subset whose restrictions form the simple roots of $\boldsymbol{a}_j$. Then $\mathrm{Re}(-s\Lambda) - \rho_j \in \boldsymbol{a}_j^*$ implies

$$(\mathrm{Re}(-s\Lambda) - \rho_j)(H_j(g)) = \sum_t b_t(\Lambda)\alpha_t(H_j(g)) ,$$

where $b_t(\Lambda)$ are real-valued continuous functions on $\boldsymbol{a}_C^*$. Therefore, there exist real-valued continuous functions $d_r(\Lambda)$ such that

$$(\mathrm{Re}(-s\Lambda) - \rho_j)(H_j(g)) = \sum_{r=1}^{m} d_r(\Lambda)\alpha_r(H_0(g)) .$$

Fix $\Lambda \in \boldsymbol{a}_C^*$ first. If $d_r(\Lambda) \geq 0$,

$$d_r(\Lambda)\alpha_r(H_0(g)) \leq d_r(\Lambda)t_0 \quad \text{for} \quad g \in S_0 ;$$

whereas if $d_r(\Lambda) < 0$,

$$d_r(\Lambda)\alpha_r(H_0(g)) \le d_r(\Lambda)\eta(H_0(g)) .$$

It follows that

$$|e^{(-s\Lambda-\rho_j)(\log a_j)}| \le \exp\left[|t_0| \sum_{d_r(\Lambda)\ge 0} d_r(\Lambda)\right] \exp\left[\sum_{d_r(\Lambda)<0} d_r(\Lambda)\eta(H_0(g))\right]$$

$$\le \exp\left[|t_0| \sum_{r=1}^{m} |d_r(\Lambda)|\right] \exp\left[-\sum_{r=1}^{m} |d_r(\Lambda)|\eta(H_0(g))\right]$$

if $\eta(H_0(g)) < 0$. Hence for any compact subset E of $\boldsymbol{a}_C^*$ there exist B, B′ > 0 such that for all $g \in S_0$ with $\eta(H_0(g) < 0$,

$$\left| e^{(-s\Lambda-\rho_j)(\log a_j)} \right| \le B' e^{-B\eta(H_0(g))} \qquad (\Lambda \in E) .$$

But

$$\{ g \in S_0 : \eta(H_0(g)) \ge 0 \} = \{ g \in S_0 : 0 \le \alpha_r(H_0(g)) \le t_0 \quad \text{for each } \alpha_r \in \Sigma^o \} ,$$

which is compact. So we can majorize the continuous

$$\left| e^{(-s\Lambda-\rho_j)(\log a_j)} \right| / e^{-B\eta(H_0(g))}$$

over $E \times \{ g_0 \in S_0 : \eta(H_0(g)) \ge 0 \}$. This completes the proof. □

Fix a compact subset E of $\boldsymbol{a}_C^*$. By the definition of $[I_{j,s,n}]_F$, for $(\Lambda, g) \in E \times S_0$ such that $\gamma g \in F \subset S_0$ for some $\gamma \in \Gamma$,

$$[I_{j,s,n}]_F (\Lambda, g) = I_{j,s,n}(\Lambda, \gamma g) .$$

Then

$$|[I_{j,s,n}]_F (\Lambda, g)| \le C' e^{-C\eta(H_0(\gamma g))} .$$

Let us relate $\eta(H_0(\gamma g))$ to $\eta(H_0(g))$ for $g \in S_0$ and $\gamma \in \Gamma$ such that $\gamma S_0 \cap S_0 \neq \phi$. First of all, we know from reduction theory that the set $\{\gamma \in \Gamma : \gamma S_0 \cap S_0 \neq \phi\}$ is finite. This is the *Siegel property* of Siegel domains (Borel [1, pp. 29-34, 103]). Let this set be denoted by $\{\gamma_k\}$. We recall the *Bruhat decomposition*:

$$G = \dot{\bigcup_{s \in W(\mathfrak{a}_0)}} P_0 w_s N_0$$

(Harish-Chandra [2, p. 34], Warner [1, p. 56]), where for each $s \in W(\mathfrak{a}_0)$ one fixes $w_s \in N(A_0)$ such that $\mathrm{Ad}(w_s) = s$ on $\mathfrak{a}_0$. Write

$$\gamma_k = n_k t_k w_s n_k' ,$$

where $n_k \in N_0$, $t_k \in A_0M_0$, $n_k' \in N_0$. For $g \in S_0$, $g = nak$, where $n \in N_0$, $a \in A_0$, $k \in K$, we have

$$\begin{aligned}\gamma_k g &= n_k t_k w_s n_k' nak \\ &= n_k t_k w_s a w_s^{-1}(w_s a^{-1} n_k' n a w_s^{-1}) w_s k .\end{aligned}$$

Denote $w_s a^{-1} n_k' n a w_s^{-1}$ by c and write $c = n_c a_c k_c$, where $n_c \in N_0$, $a_c \in A_0$, $k_c \in K$. Then

$$\begin{aligned}\gamma_k g &= n_k t_k w_s a w_s^{-1} n_c a_c k_c w_s k \\ &= n_k[t_k(w_s a w_s^{-1})] n_c [t_k(w_s a w_s^{-1})]^{-1} t_k(w_s a w_s^{-1}) a_c k_c w_s k \\ &= n_k[t_k(w_s a w_s^{-1})] n_c [t_k(w_s a w_s^{-1})]^{-1} t_k'(w_s a w_s^{-1}) a_c t_k'' k_c w_s k ,\end{aligned}$$

where

$$n_k[t_k(w_s a w_s^{-1})] n_c [t_k(w_s a w_s^{-1})]^{-1} \in N_0 ,$$

$$t_k'(w_s a w_s^{-1}) a_c \in A_0 ,$$

$$t_k'' k_c w_s k \in K ;$$

$$t_k = t_k' t_k'' .$$

Thus $a(\gamma_k g) = (w_s a w_s^{-1}) t_k' a_c$. So

$$H_0(\gamma_k g) = \log w_s a w_s^{-1} + \log t_k' + \log a_c .$$

For $\alpha_\tau \in \Sigma^o$,

$$\alpha_\tau(H_0(\gamma_k g)) = \alpha_\tau(\log w_s a w_s^{-1}) + \alpha_\tau(\log t_k') + \alpha_\tau(\log a_c) .$$

Since $\{k\}$ is finite, there exists a constant D_1 such that $\alpha_r(\log t_k') \geq D_1$ for all k and all r. As to $\alpha_r(\log a_c)$, where $c = w_s a^{-1} n_k' n a w_s^{-1}$, let us recall another basic result from reduction theory.

Lemma 3.4.2. If ω is a relatively compact subset of N_0, then

$$\bigcup_{a \in A_0(t_0)} a^{-1}\omega a$$

is also relatively compact in N_0.

(See Borel [1, pp. 14, 85].)

Since $n \in \Omega_{N_0}$, which is a compact subset of N_0, the lemma implies that

$$\{c\} = \bigcup_{s \in W(\mathfrak{a}_0)} \bigcup_{a \in A(t_0)} \bigcup_{k} w_s a^{-1} n_k' \Omega_{N_0} a w_s^{-1}$$

is a relatively compact subset of G. Therefore, by virtue of the continuous projection $\pi_{A_0} : N_0 A_0 K_0 \to A_0$, there exists a constant D_2 such that $\alpha_r(\log a_c) \geq D_2$ for all such group elements c and r. Thus,

$$\alpha_r(H_0(\gamma g)) \geq \alpha_r(\log w_s a w_s^{-1}) + D_3 \geq \eta(\log w_s a w_s^{-1}) + D_3$$

for some constant D_3.

The next lemma will complete the proof that $[\, I_{j,s,n} \,]_F$ is slowly increasing. However, here we limit our demonstration to the case of $G = SL(n, \mathbf{R})$.

Lemma 3.4.3. There exist constants $B, B' > 0$ such that for $a \in A(t_0)$ with $\eta(\log a) < 0$,

$$\alpha_r(\log w_s a w_s^{-1}) \geq B\eta(\log a) - B'$$

for all $s \in W(\mathfrak{a}_0)$ and all r.

Proof. It is known that

$$a_0 = \left\{ \begin{bmatrix} x_1 & & & \\ & \cdot & & \\ & & \cdot & \\ & & & \cdot \\ & & & & x_n \end{bmatrix} : x_1 + \cdots + x_n = 0 \right\}.$$

For $s \in W(a_0)$,

$$w_s a w_s^{-1} = \begin{bmatrix} x_{\sigma(1)} & & & \\ & \cdot & & \\ & & \cdot & \\ & & & \cdot \\ & & & & x_{\sigma(n)} \end{bmatrix}$$

for some $\sigma \in S_n$, the permutation group on n letters. Hence,

$$\alpha_r(\log w_s a w_s^{-1}) = x_{\sigma(r)} - x_{\sigma(r+1)} \cdot$$

Either $\sigma(r) < \sigma(r+1)$ or $\sigma(r) > \sigma(r+1)$. In the first case,

$$\begin{aligned} x_{\sigma(r)} - x_{\sigma(r+1)} &= (x_{\sigma(r)} = x_{\sigma(r)+1}) + \cdots + (x_{\sigma(r+1)-1} - x_{\sigma(r+1)}) \\ &\geq (\sigma(r+1) - \sigma(r))\eta(\log a) \\ &\geq n\eta(\log a), \end{aligned}$$

since $\eta(\log a) < 0$. If $\sigma(r) > \sigma(r+1)$, then

$$\begin{aligned} x_{\sigma(r)} - x_{\sigma(r+1)} &= (x_{\sigma(r)} - x_{\sigma(r)-1}) + \cdots + (x_{\sigma(r)+1} - x_{\sigma(r+1)}) \\ &\geq -(\sigma(r) - \sigma(r+1))t_0 \\ &\geq -(\sigma(r) - \sigma(r+1))\,|\,t_0\,| \geq -n\,|\,t_0\,| . \end{aligned}$$

Thus for all r and all $s \in W(a_0)$,

$$\alpha_r(\log w_s a w_s^{-1}) \geq n\eta(\log a) - n\,|\,t_0\,| . \quad \square$$

By Lemma 3.4.3 and that which precedes it, there exist constants B, D such that for $g \in S_0$ with $\eta(H_0(g)) < 0$,

$$\eta(H_0(\gamma g)) \geq B\eta(H_0(g)) + D\,.$$

Hence,

$$|\,[\,I_{j,s,n}\,]_F(\Lambda, g)\,| \leq C'e^{-CB\eta(H_0(g))-CD} = C'e^{-CB\eta(H_0(g))}$$

for $\Lambda \in E$ and all $g \in S_0$ with $\eta(H_0(g)) < 0$. But $\{\, g \in S_0 : \eta(H_0(g)) \geq 0 \,\}$ is compact, and $[\,I_{j,s,n}\,]_F(\Lambda, g)$ is bounded on every compact subset of $\boldsymbol{a}_C^* \times G$, so

$$|\,[\,I_{j,s,n}\,]_F(\Lambda, g)\,| / e^{-CB\eta(H_0(g))}$$

can be majorized over $E \times \{\, g \in S_0 : \eta(H_0(g)) \geq 0 \,\}$.

We have therefore proved that $[\,I_{j,s,n}\,]_F$ is slowly increasing. This implies that $H^o_{j,s,n}$ is slowly increasing also. Indeed we have proved a little more, as we have all along allowed Λ to vary over E, a compact subset of $\boldsymbol{a}_C^*$: there exist $C_1, C_2 > 0$ such that

$$|H^o_{j,s,n}(\Lambda, g)| \leq C_1 e^{-C_2\eta(H_0(g))} \qquad \text{for } (\Lambda, g) \in E \times S_0\,.$$

The main result of this section is now within our reach.

<u>Theorem 3.4.2.</u> For every compact subset E of $\boldsymbol{a}_C^*$, there exists $\kappa > 0$ such that

$$|\,\mathbf{K}_\alpha H^o_{j,s,n}(\Lambda, g)\,| \leq \kappa \quad \text{for } (\Lambda, g) \in E \times G\,,$$

and hence

$$\mathbf{K}_\alpha H^o_{j,s,n}(\Lambda, g) \in L^2(\Gamma \backslash G) \quad \text{for each } \Lambda \in \boldsymbol{a}_C^*\,.$$

Moreover,

$$\mathbf{K}_\alpha H^o_{j,s,n}(\Lambda, g) \in C^\infty(\boldsymbol{a}_C^* \times G)$$

and is holomorphic on $\boldsymbol{a}_C^*$ for each $g \in G$.

Proof. Lemma 2.4.2(ii) implies the first part of the theorem, while the second part follows from Theorem 3.4.1, Proposition 3.3.3 and the remark after it. □

3.5. Truncation of the Eisenstein Series

The crucial growth property of the Eisenstein series is stated in Lemma 3.5.1.

Lemma 3.5.1. For each $\Lambda \in (\boldsymbol{a}_C^*)^+$, the Eisenstein series $E(\Lambda, g)$ is a slowly increasing function.

(See Harish-Chandra [2, pp. 7, 30].)

We may now examine the effect the compact operator $\mathbf{K}_\alpha$ has on the Eisenstein series. From the definition of $\mathbf{K}_\alpha$,

$$\mathbf{K}_\alpha E = \lambda_\alpha(\mathbf{K}_\alpha^o E) = \lambda_\alpha\left[\alpha * E - \sum_{r=1}^{m} \delta_r\alpha * E^r\right]_F$$

$$= \lambda_\alpha\left[\alpha * E - \sum_{r=1}^{m} [\delta_r\alpha * E^r]_F\right]$$

$$= \lambda_\alpha(\alpha * E) - \sum_{r=1}^{m} \lambda_\alpha[\delta_r\alpha * E^r]_F .$$

Now

$$\alpha * E^r(\Lambda, g) = \int_G \alpha(h^{-1}g) \int_{N_r \cap \Gamma \setminus N_r} |E(\Lambda, nh)|\, dndh .$$

Note that

$$\int_G |\alpha(h^{-1}g)| \int_{N_r \cap \Gamma \setminus N_r} |E(\Lambda, nh)|\, dndh < \infty$$

because α is of compact support, hence Fubini's theorem enables us to write

$$\int_G \alpha(h^{-1}g) \int_{N_r \cap \Gamma \backslash N_r} E(\Lambda, nh)\, dndh = \int_{N_r \cap \Gamma \backslash N_r} \int_G \alpha(h^{-1}g)E(\Lambda, nh)\, dhdn$$

$$= \int_{N_r \cap \Gamma \backslash N_r} \int_G \alpha(h^{-1}ng)E(\Lambda, h)\, dhdn$$

$$= \int_{N_r \cap \Gamma \backslash N_r} \hat{\alpha}^o(\Lambda)E(\Lambda, ng)\, dn = \hat{\alpha}^o(\Lambda)E^r(\Lambda, g)\,.$$

Of course,

$$\lambda_\alpha(\,\alpha * E\,) = \alpha * \alpha * \alpha * E = \hat{\alpha}(\Lambda)E\,,$$

where $\hat{\alpha}(\Lambda) = (\hat{\alpha}^o(\Lambda))^3$. Therefore,

$$\mathbf{K}_\alpha E(\Lambda, g) = \hat{\alpha}(\Lambda)E(n, g) - \sum_{r=1}^{m} \hat{\alpha}^o(\Lambda)\lambda_\alpha[\,\delta_r E^r\,]_F(\Lambda, g)$$

$$= \hat{\alpha}(\Lambda)E(\Lambda, g) - \sum_j \hat{\alpha}^o(\Lambda)\lambda_\alpha[\,\delta_j E^j\,]_F(\Lambda, g)$$

$$= \hat{\alpha}(\Lambda)E(\Lambda, g) - \hat{\alpha}(\Lambda)\sum_j \frac{\lambda_\alpha[\,\delta_j E^j\,]_F(\Lambda, g)}{(\hat{\alpha}^o(\Lambda))^2}$$

if $\Lambda \in (\boldsymbol{a}_C^*)^+ \setminus \hat{\alpha}^{-1}(\{0\})$.

In view of Theorem 2.2.2, we have the following result.

<u>Theorem 3.5.1.</u> For each $\Lambda \in (\boldsymbol{a}_C^*)^+ \setminus \hat{\alpha}^{-1}(\{0\})$,

$$E(\Lambda, g) - \frac{1}{(\hat{\alpha}^o(\Lambda))^2}\sum_j \lambda_\alpha[\,\delta_j E^j\,]_F(n, g)$$

belongs to $L^2(\Gamma \setminus G)$.

Of course,

$$E(\Lambda, g) - \frac{1}{(\hat{\alpha}^o(\Lambda))^2}\sum_j \lambda_\alpha[\,\delta_j E^j\,]_F(\Lambda, g)$$

is actually bounded on G. More basically, using $\mathbf{K}_\alpha^o$ in place of $\mathbf{K}_\alpha$, we obtain from Lemma

2.4.2(ii) that

$$E(\Lambda, g) - \sum_j [\, \delta_j E^j \,]_F (\Lambda, g)$$

is bounded on G. This is only one incident of a general phenomenon. For later reference we make explicit a more general result, which is a consequence of Theorem 2.2.2 and Lemma 2.4.2(ii).

Theorem 3.5.2. If f is Γ-automorphic, slowly increasing, K-finite and **Z**(G)-finite, then

$$f - \sum_{r=1}^{m} [\, \delta_r f^r \,]_F$$

is bounded.

We have seen that the Eisenstein series satisfies the following equation:

$$(\, \mathbf{K}_\alpha - \hat{\alpha}(\Lambda)\,)E = -\sum_j \hat{\alpha}^o(\Lambda)\lambda_\alpha[\, \delta_j E^j \,]_F \qquad (\, \Lambda \in (\mathfrak{a}_C^*)^+ \,)\,.$$

Since

$$E^j(\Lambda, g) = \sum_{s,n} \phi(j, s, n \mid \Lambda) e^{(-s\Lambda - \rho_j)(\log a_j)} \sigma(k)\Psi_{j,s,n}(m_j)\,,$$

$$[\, \delta_j E^j \,]_F (\Lambda, g) = \sum_{s,n} \phi(j, s, n \mid \Lambda)[\, \delta_j e^{(-s\Lambda - \rho_j)(\log a_j)} \sigma(k)\Psi_{j,s,n}(m_j) \,]_F$$

$$= \sum_{s,n} \phi(j, s, n \mid \Lambda)[\, I_{j,s,n} \,]_F (\Lambda, g)\,.$$

Thus

$$(\, \mathbf{K}_\alpha - \hat{\alpha}(\Lambda)\,)E = -\sum_{j,s,n} \phi(j, s, n \mid \Lambda)\hat{\alpha}^o(\Lambda)\lambda_\alpha[\, I_{j,s,n} \,]_F$$

$$= -\sum_{j,s,n} \phi(j, s, n \mid \Lambda)\hat{\alpha}^o(\Lambda) H^o_{j,s,n} \qquad (\, \Lambda \in (\mathfrak{a}_C^*)^+ \,)\,.$$

Hereafter, to simplify the notations, we shall often use a multi-index J in place of (j, s, n). For

$\Lambda \in (\boldsymbol{a}_{\mathrm{C}}^{*})^{+} \setminus \hat{\alpha}^{-1}(\{0\})$, we then have

$$(3.5.1) \qquad (\,\mathbf{K}_{\alpha} - \hat{\alpha}(\Lambda)\,)E(\Lambda) = -\sum_{j} \phi(J \mid \Lambda)\hat{\alpha}(\Lambda)H_J(\Lambda)$$

where $\hat{\alpha} = (\hat{\alpha}^{o})^{3}$ and

$$H_J = \frac{H_J^{o}}{(\hat{\alpha}^{o})^{2}}\ .$$

We now present a set of Fredholm operations, whose solutions combine to give a representation for the Eisenstein series that is at the heart of this development for the analytic continuation of Eisenstein series.

The *Fredholm equations* are as follows: for all J,

$$(3.5.2) \qquad (\,\mathbf{K}_{\alpha} - \hat{\alpha}(\Lambda)\,)\,{}^{o}F_J^{**}(\Lambda) = -\,\mathbf{K}_{\alpha}H_J^{o}(\Lambda) \qquad (\,\Lambda \in \boldsymbol{a}_{\mathrm{C}}^{*} \setminus \hat{\alpha}^{-1}(\mathrm{spec}\ \mathbf{K}_{\alpha})\,)\,.$$

Since $-\,\mathbf{K}_{\alpha}H_J^{o}(\Lambda) \in L^2(\Gamma \setminus G)$ by Theorem 3.4.2, for $\Lambda \in \boldsymbol{a}_{\mathrm{C}}^{*} \setminus \hat{\alpha}^{-1}(\mathrm{spec}\ \mathbf{K}_{\alpha})$, there exist ${}^{o}F_J^{**}(\Lambda) \in L^2(\Gamma \setminus G)$ which are solutions of the equation (3.5.2). As $\mathbf{K}_{\alpha}$ is compact, $0 \in \mathrm{spec}\ \mathbf{K}_{\alpha}$ (Rudin [1, p. 99]), so

$$\Lambda \in \boldsymbol{a}_{\mathrm{C}}^{*} \setminus \hat{\alpha}^{-1}(\mathrm{spec}\ \mathbf{K}_{\alpha}) \ \text{ implies } \ \hat{\alpha}(\Lambda),\, \hat{\alpha}^{o}(\Lambda) \neq 0\,.$$

Set

$$F_J^{**}(\Lambda) = \frac{{}^{o}F_J^{**}(\Lambda)}{(\hat{\alpha}^{o}(\Lambda))^{2}}\ ,$$

then the equation (3.5.2) becomes

$$(\,\mathbf{K}_{\alpha} - \hat{\alpha}(\Lambda)\,)F_J^{**}(\Lambda) = -\,\mathbf{K}_{\alpha}H_J(\Lambda) \qquad (\,\Lambda \in \boldsymbol{a}_{\mathrm{C}}^{*} \setminus \hat{\alpha}^{-1}(\mathrm{spec}\ \mathbf{K}_{\alpha})\,)\,.$$

Now set

$$(3.5.3) \qquad F_J^{*}(\Lambda) = F_J^{**}(\Lambda) + H_J(\Lambda)$$

and

$$F(\Lambda) = \sum_J \phi(J \mid \Lambda) F_J^*(\Lambda) ,$$

where $\Lambda \in (\boldsymbol{a}_C^*)^+ \setminus \hat{\alpha}^{-1}(\text{spec } \mathbf{K}_\alpha)$. Then

$$(\mathbf{K}_\alpha - \hat{\alpha}(\Lambda))F = \sum_J \phi(J \mid \Lambda)[(\mathbf{K}_\alpha - \hat{\alpha}(\Lambda))F_J^{**} + (\mathbf{K}_\alpha - \hat{\alpha}(\Lambda))H_J]$$

$$= \sum_J \phi(J \mid \Lambda)[- \mathbf{K}_\alpha H_J + (\mathbf{K}_\alpha - \hat{\alpha}(\Lambda))H_J]$$

$$= - \sum_J \phi(J \mid \Lambda)\hat{\alpha}(\Lambda)H_J .$$

So $F(\Lambda)$ satisfies the equation (3.5.1) like $E(\Lambda)$ does. Of course, for $F(\Lambda)$, $\Lambda \in \boldsymbol{a}_C^* \setminus \hat{\alpha}^{-1}(\text{spec}\mathbf{K}_\alpha)$; whereas for $E(\Lambda)$, $\Lambda \in (\boldsymbol{a}_C^*)^+ \setminus \hat{\alpha}^{-1}(\{0\})$. By its construction,

$$F(\Lambda) - \sum_J \phi(J \mid \Lambda)H_J(\Lambda) \quad \in L^2(\Gamma \setminus G) .$$

The same can be said of

$$E(\Lambda) - \sum_J \phi(J \mid \Lambda)H_J(\Lambda)$$

by virtue of Theorem 3.5.1. Therefore, for $\Lambda \in \boldsymbol{a}_C^* \setminus \hat{\alpha}^{-1}(\text{spec}\mathbf{K}_\alpha)$,

$$E(\Lambda) = F(\Lambda) = \sum_J \phi(J \mid \Lambda)F_J^*(\Lambda) = \sum_{j,s,n} \phi(j, s, n \mid \Lambda)F_{j,s,n}^*(\Lambda) . \tag{3.5.4}$$

Referring back to Theorem 2.2.1 and Proposition 1.4.4, we know that the kernel function α can be chosen so that $\hat{\alpha}^o(\Lambda)$, hence $\hat{\alpha}(\Lambda)$ also, is a nonconstant entire function on $\boldsymbol{a}_C^*$.

3.6. Holomorphicity of Fredholm Solutions

Recall the basic Fredholm equations (3.5.2) :

$$(\mathbf{K}_\alpha - \hat{\alpha}(\Lambda))^o F_J^{**}(\Lambda) = - \mathbf{K}_\alpha H_J^o(\Lambda) \qquad (\Lambda \in \boldsymbol{a}_C^* \setminus \hat{\alpha}^{-1}(\text{spec } \mathbf{K}_\alpha)) ,$$

where $\mathbf{K}_\alpha : L^2(\Gamma \setminus G) \to L^2(\Gamma \setminus G)$. In this section we shall be concerned with the smoothness of each Fredholm solution ${}^{o}F_J^{**}(\Lambda, g)$ with respect to Λ and g, in particular its holomorphicity with respect to Λ.

We know that $\mathbf{K}_\alpha$ is a compact operator on the Banach space $L^2(\Gamma \setminus G)$. The standard theory of compact operators then asserts that spec $\mathbf{K}_\alpha$ is a countable closed set which has no limit point except possibly zero (Rudin [1, p. 103]), and that every nonzero element of spec $\mathbf{K}_\alpha$ is a pole of the *resolvent* $R(\lambda)$ (Yosida [1, pp. 211, 230, 285]). Since $\hat{\alpha}$ is a non-constant entire function, it follows that $\hat{\alpha}^{-1}(\text{spec } \mathbf{K}_\alpha)$ is a countable closed set with a discrete set of limit points, all of which correspond to the point 0. At each isolated singularity λ_0, $R(\lambda)$ has a Laurent expansion:

$$R(\lambda) = \sum_{n=-p}^{\infty} (\lambda - \lambda_0)^n A_n ,$$

where A_n are linear operators on $L^2(\Gamma \setminus G)$ given by

$$A_n = (2\pi i)^{-1} \int_C (\nu - \lambda_0)^{-n-1} R(\nu)\, d\nu ,$$

C being a sufficiently small positively oriented circle centered at λ_0 (see Yosida [1, p. 228]). Of course, $p = 0$ if λ_0 is a regular point or a removable singularity.

Let us fix a nonzero element of $\mathfrak{a}_\mathbf{C}^*$ to form a basis and thus parametrize $\mathfrak{a}_\mathbf{C}^*$ by a complex variable. In the rest of this section, $\mathfrak{a}_\mathbf{C}^*$ will be identified with $\mathbf{C}$. In this section and in Section 4.3 we shall often use the notion of a deleted neighborhood. A *deleted neighborhood* of $z \in \mathfrak{a}_\mathbf{C}^*$ is a neighborhood of z with z itself removed. If $z_0 \in \hat{\alpha}^{-1}(\{\lambda_0\})$, A_n can be expressed as a path integral in $\mathfrak{a}_\mathbf{C}^*$:

$$A_n = (2\pi i)^{-1} \int_\gamma (\hat{\alpha}(w) - \hat{\alpha}(z_0))^{-n-1} \hat{\alpha}'(w) R(\hat{\alpha}(w))\, dw .$$

Then, in a deleted neighborhood U of z_0,

$$(3.6.1) \qquad {}^{o}F_J^{**}(z) = \sum_{n=-p}^{\infty} (\hat{\alpha}(z) - \hat{\alpha}(z_0))^n A_n\{-\mathbf{K}_\alpha H_J^o(z)\} \qquad (z \in U)$$

where

$$A_n\{-\mathbf{K}_\alpha H_J^o(z)\} = (2\pi i)^{-1} \int_\gamma (\hat{\alpha}(w) - \hat{\alpha}(z_0))^{-n-1}\hat{\alpha}'(w)R\,(\hat{\alpha}(w))\{-\mathbf{K}_\alpha H_J^o(z)\}\,dw\,. \tag{3.6.2}$$

One is inclined to interpret equations (3.6.1) and (3.6.2) as more than L^2-statements, but this calls for verification.

First we introduce some notations for Proposition 3.6.1. Let X be a topological measure space, $L^p(X)$ the corresponding L^p-space for some $p \geq 1$, and $B(L^p(X))$ the space of bounded linear operators on $L^p(X)$.

Proposition 3.6.1. Let $\gamma : [t_1, t_2] \to \mathbf{C}$ be a path in the complex plane, and T a continuous map from γ^*, the range of γ, into $B(L^p(X))$. If $f \in L^p(X)$ is such that as w varies over γ^* representative of the L^p-class $T(w)f$ can be chosen so that $(T(w)f))(x)$ is continuous in w for each fixed $x \in X$, then

$$\left[\int_\gamma T(w)\,dw\right] f = \int_\gamma T(w)f\,dw = F\,,$$

where F is the L^p-class represented by the function:

$$\int_\gamma (T(w)f)(x)\,dw \qquad\qquad (x \in X)\,.$$

Proof. The first equality follows from the definition of a path integral of a Banach space valued function, and is true for all $f \in L^p(X)$. To verify the second equality, for each positive integer n, let

$$t_1 < t_1 + \frac{1}{n}(t_2 - t_1) < t_1 + \frac{2}{n}(t_2 - t_1) < \cdots < t_2$$

be a partition of $[t_1, t_2]$. From the definition of a path integral,

$$\lim_{n\to\infty} \sum_{i=0}^{n-1} \frac{1}{n}\,\gamma(t_1 + \frac{i}{n} + \frac{1}{2n})T(\gamma(t_1 + \frac{i}{n} + \frac{1}{2n}))f = \int_\gamma T(w)f\,dw\,.$$

Now fix a representative of the L^p-class $\int_\gamma T(w)f\,dw$. If as w varies of γ^* a particular representation of T(w)f is chosen, then from the *Riesz-Fischer theorem* it is true that for almost all x,

$$\left[\int_\gamma T(w)f\,dw\right](x) = \lim_{n\to\infty} \sum_{i=0}^{n-1} \frac{1}{n}\gamma'(t_1 + \frac{i}{n} + \frac{1}{2n})T(\gamma(t_1 + \frac{i}{n} + \frac{1}{2n}))f(s)\,.$$

On the other hand, if for each x, (T(w)f)(x) is continuous in w, the right-hand side of the last equation is just $\int_\gamma (T(w)f)(x)\,dw$. Thus, with this choice of representative of $\int_\gamma T(w)f\,dw$ as w varies over γ^*,

$$\left[\int_\gamma T(w)f\,dw\right](x) = \int_\gamma (T(w)f)(x)\,dw \qquad \text{for almost all } x\,. \quad \square$$

Let us now return to the original situation.

<u>Lemma 3.6.1.</u> There exists $\kappa > 0$ such that for all $f \in L^2(\Gamma \backslash G)$,

$$|\mathbf{K}_\alpha^m f(g)| \le \kappa^m \mathrm{Vol}(\Gamma\backslash G)^{(m-1)/2}\, \|f\|_2 \qquad (m = 1, 2, \cdots)\,.$$

<u>Proof.</u> By Proposition 2.4.2,

$$|\mathbf{K}_\alpha f(g)| \le \kappa \|f\|_2 \qquad (g \in G)\,,$$

so

$$\|\mathbf{K}_\alpha f\| \le \kappa \|f\|\,\mathrm{Vol}(\Gamma\backslash G)^{1/2}\,.$$

Similarly,

$$|\mathbf{K}_\alpha^2 f(g)| \le \kappa \|\mathbf{K}_\alpha f\|_2 \le \kappa^2 \|f\|_2 \mathrm{Vol}(\Gamma\backslash G)^{1/2}\,.$$

The general assertion can be obtained by induction. $\square$

Lemma 3.6.2. Let f be a bounded continuous function in $L^2(\Gamma \backslash G)$. As w varies over $\mathfrak{a}_C^* \backslash \hat{\alpha}^{-1}(\mathrm{spec}\,\mathbf{K}_\alpha)$ representative of $R(\hat{\alpha}(w))f$ can be chosen so that $R(\hat{\alpha}(w))f(g)$ is a continuous function on $\mathfrak{a}_C^* \backslash \hat{\alpha}^{-1}(\mathrm{spec}\,\mathbf{K}_\alpha) \times G$.

Proof. Let $w \in \mathfrak{a}_C^* \backslash \hat{\alpha}^{-1}(\mathrm{spec}\mathbf{K}_\alpha)$ and N be the inverse image under $\hat{\alpha}$ of an open disc centered at $\hat{\alpha}(w)$ that avoids all points of spec $\mathbf{K}_\alpha$. Then

$$R(\hat{\alpha}(z)) = \sum_{n=0}^{\infty} (\hat{\alpha}(z) - \hat{\alpha}(w))^n R(\hat{\alpha}(w))^{n+1} \qquad (z \in N).$$

(See Yosida [1, p. 211].) Note that $\hat{\alpha}(w) \neq 0$ because $0 \in \mathrm{spec}\,\mathbf{K}_\alpha$. Hence, as a L^2-class in $L^2(\Gamma \backslash G)$,

$$R(\hat{\alpha}(z))f = \sum_{n=0}^{\infty} (\hat{\alpha}(z) - \hat{\alpha}(w))^n R(\hat{\alpha}(w))^{n+1} f \qquad (z \in N).$$

Denote $R(\hat{\alpha}(w))^{n+1}f$ by f_n. Then

$$R(\hat{\alpha}(z))f = \sum_{n=0}^{\infty} (\hat{\alpha}(z) - \hat{\alpha}(w))^n f_n \qquad (z \in N).$$

It is clear that $f_n \in L^2(\Gamma \backslash G)$,

$$\| f_n \|_2 \leq \| R(\hat{\alpha}(w)) \|^{n+1} \| f \|_2 ,$$

and

$$(\mathbf{K}_\alpha - \hat{\alpha}(w))^{n+1} f_n = f .$$

Now by the binomial theorem

$$(\mathbf{K}_\alpha - \hat{\alpha}(w))^{n+1} f_n = \sum_{k=0}^{n} \binom{n+1}{k} \mathbf{K}_\alpha^{n+1-k}(-\hat{\alpha}(w))^k f_n + (-\hat{\alpha}(w))^{n+1} f_n .$$

Therefore,

$$f_n = (-\hat{\alpha}(w))^{-(n+1)} \left[f - \sum_{k=0}^{n} \binom{n+1}{k} (-\hat{\alpha}(w))^k \mathbf{K}_\alpha^{n+1-k} f_n \right] ,$$

If we begin with a particular representative f_n, the right-hand side of the last equation will also be a representative. By Proposition 3.3.3, this is a continuous function in $L^2(\Gamma \setminus G)$. Besides, it can easily be majorized over G:

$$|f_n(g)| = \left| (-\hat{\alpha}(w))^{-(n+1)} \left[f(g) - \sum_{k=0}^{n} \binom{n+1}{k} (-\hat{\alpha}(w))^k \mathbf{K}_\alpha^{n+1-k} f_n(g) \right] \right|$$

$$\leq |\hat{\alpha}(w)|^{-(n+1)} \left[|f(g)| + \sum_{k=0}^{n} \binom{n+1}{k} |\hat{\alpha}(w)|^k |\mathbf{K}_\alpha^{n+1-k} f_n(g)| \right],$$

and by Lemma 3.6.1,

$$\leq |\hat{\alpha}(w)|^{-(n+1)} \left[|f(g)| + \sum_{k=0}^{n} \binom{n+1}{k} |\hat{\alpha}(w)|^k \kappa^{n+1-k} \mathrm{Vol}(\Gamma \setminus G)^{(n+1-k-1)/2} \, \|f_n\|_2 \right]$$

$$\leq |\hat{\alpha}(w)|^{-(n+1)} \left[|f(g)| + \sum_{k=0}^{n} \binom{n+1}{k} |\hat{\alpha}(w)|^k \kappa^{n+1-k} \mathrm{Vol}(\Gamma \setminus G)^{(n+1-k-1)/2} \, \|R(\hat{\alpha}(w))\|^{n+1} \|f\|_2 \right]$$

$$\leq |\hat{\alpha}(w)|^{-(n+1)} [\, |f(g)| + \mathrm{Vol}(\Gamma \setminus G)^{-1/2} \|f\|_2 \|R(\hat{\alpha}(w))\|^{n+1}$$

$$\times \sum_{k=0}^{n} \binom{n+1}{k} |\hat{\alpha}(w)|^k (\kappa \, \mathrm{Vol}(\Gamma \setminus G)^{1/2})^{n+1-k}]$$

$$\leq |\hat{\alpha}(w)|^{-(n+1)} \left[C_1 + C_2 \|R(\hat{\alpha}(w))\|^{n+1} \sum_{k=0}^{n} \binom{n+1}{k} |\hat{\alpha}(w)|^k (\kappa \, \mathrm{Vol}(\Gamma \setminus G)^{1/2})^{n+1-k} \right]$$

$$\leq C_1 \left(\frac{1}{|\hat{\alpha}(w)|} \right)^{n+1} + C_2 \left[\frac{\|R(\hat{\alpha}(w))\| (|\hat{\alpha}(w)| + \kappa \, \mathrm{Vol}(\Gamma \setminus G)^{1/2})}{|\hat{\alpha}(w)|} \right]^{n+1}$$

Henceforth f_n denotes this representation.

If now a particular representative of $R(\hat{\alpha}(z))f$ has been chosen, then for almost all $g \in \Gamma \setminus G$,

$$R(\hat{\alpha}(z))f(g) = \lim_{p \to \infty} \sum_{n=0}^{p} (\hat{\alpha}(z) - \hat{\alpha}(w))^n f_n(g) \qquad (z \in N).$$

On the other hand, the preceding development implies that there is a neighborhood $\tilde{N}$ of w, $\tilde{N} \subset N$, such that

$$\sum_{n=0}^{\infty} (\hat{\alpha}(z) - \hat{\alpha}(w))^n f_n(g)$$

converges uniformly on $\tilde{N} \times G$. As z varies over $\tilde{N}$, we can let $R(\hat{\alpha}(z))f(g)$ be

$$\sum_{n=0}^{\infty} (\hat{\alpha}(z) - \hat{\alpha}(w))^n f_n(g) \qquad (g \in \Gamma \backslash G).$$

For each $w \in \boldsymbol{a}_C^* \backslash \hat{\alpha}^{-1}(\text{spec } \mathbf{K}_\alpha)$, this selection process can be carried out. No inconsistency can arise because $R(\hat{\alpha}(z))f$ will always be a continuous function in $L^2(\Gamma \backslash G)$, and for a given L^2-class there can at most be one. □

In view of Proposition 3.6.1 and Lemma 3.6.2, we can for each $z \in \mathbf{C}$ take $A_n\{-\mathbf{K}H_J(z)\}(g)$ to be:

$$(2\pi i)^{-1} \int_\gamma (\hat{\alpha}(w) - \hat{\alpha}(z_0))^{-n-1} \hat{\alpha}'(w) R(\hat{\alpha}(w))\{-\mathbf{K}_\alpha H_J^o(z)\}(g)\, dw \qquad (g \in G).$$

(We shall also use the more explicit notations of $A_n\{-\mathbf{K}_\alpha H_J^o\}(z, g)$ and $R(\hat{\alpha}(w))\{-\mathbf{K}_\alpha H_J^o\}(z, g)$.)

Lemma 3.6.3. Let $w \in \boldsymbol{a}_C^* \backslash \hat{\alpha}^{-1}(\text{spec } \mathbf{K}_\alpha)$ be fixed. For each $g \in G$, $R(\hat{\alpha}(w))\{-\mathbf{K}_\alpha H_J^o\}(z, g)$ is holomorphic in z over the entire $\boldsymbol{a}_C^*$. Moreover, $R(\hat{\alpha}(w))\{-\mathbf{K}_\alpha H_J^o\}(z, g)$ belongs to $C(\boldsymbol{a}_C^* \times G)$.

Proof. By Theorem 3.4.2, for each $g \in G$, $\{-\mathbf{K}_\alpha H_J^o(z)\}(g)$ is holomorphic in z over the entire $\boldsymbol{a}_C^*$. So

$$\{-\mathbf{K}_\alpha H_J^o(z)\}(g) = \sum_{m=0}^{\infty} z^m a_m(g) \qquad (z \in \boldsymbol{a}_C^*),$$

where

$$a_m(g) = (2\pi i)^{-1} \int_\beta u^{-m-1}\{-\mathbf{K}_\alpha H_J^o(u)\}(g)\, du\,,$$

β being a positively oriented circle of radius r centered at the origin. By Theorem 3.4.2, $\{-\mathbf{K}_\alpha H_J^o(z)\}(g)$ can be majorized over $\beta^* \times G$, so there exists a constant function M(r) on G such that

$$|a_m(g)| \le \frac{M(r)}{r^m} \qquad (g \in G) \qquad \text{for all } r \ge 0 \text{ and all } m\,.$$

It follows that for each $z \in \boldsymbol{a}_C^*$, $\sum_{m=0}^{\infty} z^m a_m(g)$ converges uniformly on G. It is easily verified that the automorphicity of $\{-\mathbf{K}_\alpha H_J^o(z)\}(g)$ implies that of $a_m(g)$ for all m. Hence $a_m \in L^2(\Gamma \setminus G)$; in addition, they are continuous by virtue of Theorem 3.4.2. Therefore, for each $z \in \boldsymbol{a}_C^*$, even as a L^2-class in $L^2(\Gamma \setminus G)$,

$$-\mathbf{K}_\alpha H_J^o(z) = \lim_{p \to \infty} \sum_{m=0}^{p} z^m a_m$$

Then, because $R\,(\hat{\alpha}(w))$ is a continuous operator,

$$R\,(\hat{\alpha}(w))\{-\mathbf{K}_\alpha H_J^o(z)\} = \sum_{m=0}^{\infty} z^m R\,(\hat{\alpha}(w)) a_m\,.$$

Denote $R\,(\hat{\alpha}(w))a_m$ by ζ_m. By Lemma 3.6.2, $\zeta_m(w, g)$ can be chosen to be a continuous function on $\boldsymbol{a}_C^* \setminus \hat{\alpha}^{-1}(\text{spec } \mathbf{K}_\alpha) \times G$. Since $(\mathbf{K}_\alpha - \hat{\alpha}(w))\zeta_m = a_m$, we have

$$\zeta_m = \frac{1}{\hat{\alpha}(w)}\,[-a_m + \mathbf{K}_\alpha \zeta_m]\,.$$

Each ζ_m can therefore be chosen to be a continuous function on $\boldsymbol{a}_C^* \setminus \hat{\alpha}^{-1}(\text{spec } \mathbf{K}_\alpha) \times G$, which is majorized as follows:

$$\left| \frac{1}{\hat{\alpha}(w)}\,[-a_m(g) + \mathbf{K}_a \zeta_m(g)] \right|$$

$$\le \frac{1}{|\hat{\alpha}(w)|}\,[\,|a_m(g)| + \kappa \|\zeta_m\|_2\,]$$

$$\le \frac{1}{|\hat{\alpha}(w)|}\,[\,|a_m(g)| + \kappa \|R(\hat{\alpha}(w))\|\,\|a_m\|_2\,]$$

$$\le \frac{1}{|\hat{\alpha}(w)|}\left[\frac{M(r)}{r^m} + \kappa \|R(\hat{\alpha}(w))\|\,\frac{M(r)}{r^m}\,\mathrm{Vol}(\Gamma\backslash G)^{1/2}\right]$$

$$= \frac{M(r) + \kappa\|R(\hat{\alpha}(w))\|M(r)\,\mathrm{Vol}(\Gamma\backslash G)^{1/2}}{|\hat{\alpha}(w)|}\,\frac{1}{r^m} = \frac{M(r,w)}{r^m} \qquad (g \in G)$$

for all $r > 0$ and all m, where

$$M(r,w) = \frac{M(r) + \kappa\|R(\hat{\alpha}(w))\|M(r)\,\mathrm{Vol}(\Gamma\backslash G)^{1/2}}{|\hat{\alpha}(w)|}.$$

Now let $\zeta_m(w, g)$ denote specifically this continuous function. Then $\sum_{m=0}^{\infty} z^m \zeta_m(w, g)$ converges uniformly for $(z, w, g) \in E_1 \times E_2 \times G$, where E_1 and E_2 are arbitrary compact subsets of $\mathfrak{a}_C^*$ and $\mathfrak{a}_C^* \backslash \hat{\alpha}^{-1}(\mathrm{spec}\,\mathbf{K}_\alpha)$ respectively. Thus it defines a continuous function on $\mathfrak{a}_C^* \times \mathfrak{a}_C^* \backslash \hat{\alpha}^{-1}(\mathrm{spec}\,\mathbf{K}_\alpha) \times G$ which is automorphic in g. Because of the uniqueness mentioned in the proof of Lemma 3.6.2, we know that

$$R(\hat{\alpha}(w))\{-\mathbf{K}_\alpha H_J^o\}(z, g) = \sum_{m=0}^{\infty} z^m \zeta_m(w, g) \quad \text{for all } z \in \mathfrak{a}_C^*,\ w \in \mathfrak{a}_C^* \backslash \hat{\alpha}^{-1}(\mathrm{spec}\,\mathbf{K}_\alpha),\ \ g \in G.$$

With this infinite series representation, the assertions of the lemma become obvious. □

An immediate consequence of Lemma 3.6.3 is that, for each $g \in G$, $A_n\{-\mathbf{K}_\alpha H_J^o\}(z, g)$ is holomorphic in z over the entire $\mathfrak{a}_C^*$. Moreover, it belongs to $C(\mathfrak{a}_C^* \times G)$. Explicitly

$$A_n\{-\mathbf{K}_\alpha H_J^o\}(z, g) = (2\pi i)^{-1} \int_\gamma (\hat{\alpha}(w) - \hat{\alpha}(z_0))^{-n-1} \hat{\alpha}'(w) \cdot \sum_{m=0}^{\infty} z^m \zeta_m(w, g)\, dw$$

$$= \sum_{m=0}^{\infty} z^m \left[(2\pi i)^{-1} \int_\gamma (\hat{\alpha}(w) - \hat{\alpha}(z_0))^{-n-1} \hat{\alpha}'(w) \zeta_m(w, g)\, dw\right] \qquad (z \in \mathfrak{a}_C^*,\ g \in G)$$

It is known from Lemma 3.6.3 that

$$|\zeta_m(w, g)| \le \frac{M(r, w)}{r^m} \qquad (g \in G) \qquad \text{for all } r > 0 \text{ and all } n\,,$$

where

$$M(r, w) = \frac{M(r) + \kappa \,||R(\hat{\alpha}(w))||\, M(r)\, \mathrm{Vol}(\Gamma \backslash G)^{1/2}}{|\hat{\alpha}(w)|}\,.$$

Before proving the main result of this section, we state a lemma that gives a majoration of $A_n\{-\mathbf{K}_\alpha H_j\}(z, g)$. Note that since $||R(\hat{\alpha}(w))||$ is a continuous function of w on γ^*, M(r, w) can be majorized over γ^* for any choice of γ.

Lemma 3.6.4. For every compact subset E of $\mathfrak{a}_C^*$ there exist constants $C, C_1, C_2 > 0$ such that

$$|A_n\{-\mathbf{K}_\alpha H_j^o\}(z, g)| \le CC_1^{-n} \qquad \text{for } n \ge 0\,,$$

$$\le CC_2^{-n} \qquad \text{for } 0 > n \ge -p\,.$$

Proof. Let $r > 0$ be such that $\sum_{m=0}^{\infty} z^m r^{-m}$ converges for all $z \in E$ and for some $D > 0$, $\left|\sum_{m=0}^{\infty} z^m r^{-m}\right| < D$ for all $z \in E$. Let $D_1, D_2, D_3, D_4 > 0$ be such that for $w \in \gamma^*$,

$$0 < D_1 < |\hat{\alpha}(w) - \hat{\alpha}(z_0)| < D_2, \qquad |\hat{\alpha}'(w)| < D_3, \qquad M(r, w) < D_4\,.$$

Then

$$\left|(2\pi i)^{-1}\int_\gamma (\hat{\alpha}(w) - \hat{\alpha}(z_0))^{-n-1}\hat{\alpha}'(w)\zeta_m(w, g)\, dw\right| \le (2\pi)^{-1}D_1^{-1}D_3D_4\,l(\gamma)D_1^{-n}r^{-m} \qquad \text{for } n \ge 0\,,$$

$$\le (2\pi)^{-1}D_2^{-1}D_3D_4\,l(\gamma)D_2^{-n}r^{-m} \qquad \text{for } 0 > n \ge -p\,,$$

where $l(\gamma)$ denotes the length of γ.

We can simply let

$$C = \max\{\ (2\pi)^{-1}D_1^{-1}D_3D_4 l(\gamma)D,\ (2\pi)^{-1}D_2^{-1}D_3D_4 l(\gamma)D\ \},\ C_1 = D_1,\ C_2 = D_2.\ \square$$

We can now verify the equation (3.6.1) on the pointwise level.

Theorem 3.6.1. If λ_0 is an isolated singularity of the resolvent $R(\lambda)$, and $z_0 \in \hat{\alpha}^{-1}(\{\lambda_0\})$, then there is a deleted neighborhood U of z_0 such that for every $z \in U$,

$$^{o}F_J^{**}(z, g) = \sum_{n=-p}^{\infty} (\hat{\alpha}(z) - \hat{\alpha}(z_0))^n A_n\{-\mathbf{K}_\alpha H_J^o\}(z, g) \qquad (g \in G)$$

is the unique continuous solution to the Fredholm equation:

$$(\mathbf{K}_\alpha - \hat{\alpha}(z))\ {}^{o}F_J^{**}(z) = -\mathbf{K}_\alpha H_J^o(z),$$

the convergence of the infinite series being uniform on $U \times G$.

Proof. In view of all that which has come before, we need only show that there is a deleted neighborhood W of z_0 such that the infinite series converges uniformly on $W \times G$. Let V be a relatively compact neighborhood of z_0. It is immediate from Lemma 3.6.4 that in V there is a neighborhood $\tilde{V}$ of z_0 such that

$$\sum_{n=0}^{\infty} (\hat{\alpha}(z) - \hat{\alpha}(z_0))^n A_n\{-\mathbf{K}_\alpha H_J^o\}(z, g)$$

converges uniformly on $\tilde{V} \times G$. If $\tilde{V}$ is small enough,

$$\hat{\alpha}(z) - \hat{\alpha}(z_0) \neq 0 \quad \text{for all } z \in \tilde{V} \setminus \{z_0\}.$$

We can simply let $W = \tilde{V} \setminus \{z_0\}$. $\square$

In Theorem 3.5.1, if λ_0 is a regular point or a removable singularity of $R(\lambda)$, $p = 0$ and U includes z_0. As a consequence of the theorem, for each $z \in \mathfrak{a}_C^* \setminus \hat{\alpha}^{-1}(\text{spec } \mathbf{K}_\alpha)$ there exists a continuous function in $L^2(\Gamma \setminus G)$ which satisfies the Fredholm equation:

$$(\mathbf{K}_\alpha - \hat{\alpha}(z))^{o}F_J^{**}(z) = - \mathbf{K}_\alpha H_J^o(z) \,.$$

This function is unique. From now on, for each $z \in \boldsymbol{a}_C^* \setminus \hat{\alpha}^{-1}(\mathrm{spec}\, \mathbf{K}_\alpha)$, ${}^oF_J^{**}(\Lambda)$ refers specifically to this function.

As a matter of fact, much more can be said about the smoothness and holomorphicity of the Fredholm solution ${}^oF_J^{**}(\Lambda, g)$.

Corollary 1. For each $z_0 \in \boldsymbol{a}_C^* \setminus \hat{\alpha}^{-1}(\{0\})$, there is a neighborhood U of z_0, U deleted if $z_0 \in \hat{\alpha}^{-1}(\mathrm{spec}\, \mathbf{K}_\alpha \setminus \{0\})$, such that for each $g \in G$, ${}^oF_J^{**}(z, g)$ is holomorphic on U. Furthermore, ${}^oF_J^{**}(z, g)$ has either removable singularities or poles at the points of $\hat{\alpha}^{-1}(\mathrm{spec}\, \mathbf{K}_\alpha \setminus \{0\})$. Lastly, ${}^oF_J^{**}(z, g)$ belong to $C^\infty(\boldsymbol{a}_C^* \setminus \hat{\alpha}^{-1}(\mathrm{spec}\, \mathbf{K}_\alpha) \times G)$.

Proof. The resolvent $R(\lambda)$ has only isolated singularities in $\mathbf{C} \setminus \{0\}$, hence the assertions about holomorphicity follow from the theorem.

To prove that ${}^oF_J^{**} \in C^\infty(\boldsymbol{a}_C^* \setminus \hat{\alpha}^{-1}(\mathrm{spec}\, \mathbf{K}_\alpha) \times G)$, let $z_0 \in \boldsymbol{a}_C^* \setminus \hat{\alpha}^{-1}(\mathrm{spec}\, \mathbf{K}_\alpha)$ and deduce first from the theorem that ${}^oF_J^{**} \in C(U \times G)$. We shall apply Proposition 3.3.3. The infinite series for ${}^oF_J^{**}(z, g)$ can be differentiated term by term to yield the derivatives of ${}^oF_J^{**}(z, g)$ with respect to z. If we recall that

$$A_n\{- \mathbf{K}_\alpha H_J^o\}\, (z, g) = \sum_{m=0}^{\infty} z^m \,[\, (2\pi i)^{-1} \int_\gamma (\hat{\alpha}(w) - \hat{\alpha}(z_0))^{-n-1} \hat{\alpha}'(w) \zeta_m(w, g)\, dw \,]$$

and refer to the proof of Lemma 3.6.4, we see readily that after ν times of repeated termwise differentiation with respect to z, the resulting infinite series for $(d^\nu\, {}^oF_J^{**}/dz^\nu)(z, g)$ still converges uniformly on $\tilde{U} \times G$ for some neighborhood $\tilde{U}$ of z_0. Since $z_0 \in \boldsymbol{a}_C^* \setminus \hat{\alpha}^{-1}(\mathrm{spec}\mathbf{K}_\alpha)$ is arbitrary, this proves that ${}^oF_J^{**}(z, g)$ and all its derivatives with respect to z belong to $C(\boldsymbol{a}_C^* \setminus \hat{\alpha}^{-1}(\mathrm{spec}\, \mathbf{K}_\alpha) \times G)$. By Proposition 3.3.3., $\mathbf{K}_\alpha {}^oF_J^{**}(z)$ is a continuous function on G for each $z \in \boldsymbol{a}_C^* \setminus \hat{\alpha}^{-1}(\mathrm{spec}\mathbf{K}_\alpha)$. Now, for each z, both sides of the Fredholm equation:

$$(\mathbf{K}_\alpha - \hat{\alpha}(z))\,{}^{o}F_J^{**}(z) = -\,\mathbf{K}_\alpha H_J^{o}(z)$$

are continuous functions on G, so the equation is valid on the pointwise level. Then

$${}^{o}F_J^{**}(z, g) = \frac{\mathbf{K}_\alpha {}^{o}F_J^{**}(z, g) + \mathbf{K}_\alpha H_J^{o}(z, g)}{\hat{\alpha}(z)}$$

for $(z, g) \in \boldsymbol{a}_C^* \setminus \hat{\alpha}^{-1}(\operatorname{spec}\mathbf{K}_\alpha) \times G$. But, according to Proposition 3.3.3, $\mathbf{K}_\alpha {}^{o}F_J^{**}(z, g)$, like $\mathbf{K}_\alpha H_J^{o}(z, g)$, belongs to $C^\infty(\boldsymbol{a}_C^* \setminus \hat{\alpha}^{-1}(\operatorname{spec}\mathbf{K}_\alpha) \times G)$. Therefore ${}^{o}F_J^{**}(z, g)$ too belongs to $C^\infty(\boldsymbol{a}_C^* \setminus \hat{\alpha}^{-1}(\operatorname{spec}\mathbf{K}_\alpha) \times G)$. □

Corollary 2. For every compact subset E of $\boldsymbol{a}_C^* \setminus \hat{\alpha}^{-1}(\operatorname{spec}\mathbf{K}_\alpha)$, ${}^{o}F_j^{**}(z, g)$ is bounded on $E \times G$.

Proof. Let $z_0 \in E \subset \boldsymbol{a}_C^* \setminus \hat{\alpha}^{-1}(\operatorname{spec}\mathbf{K}_\alpha)$. Referring to the theorem and Lemma 3.6.4, there exist a neighborhood $\tilde{U}$ of z_0 and $M > 0$, $\tilde{U}$ sufficiently small and contained in U, such that

$$|{}^{o}F_J^{**}(z, g)| \le C \sum_{n=0}^{\infty} \left[\frac{|\hat{\alpha}(z) - \hat{\alpha}(z_0)|}{C_1} \right]^n < M \qquad (g \in G) \qquad \text{for all } z \in \tilde{U}.$$

Since E is covered by finitely many such neighborhoods, we have the conclusion. □

Insofar as the main line of development of this work is concerned, the relevant objects are really the functions $F_J^{**}(z, g)$, as defined by

$$F_J^{**}(z, g) = \frac{{}^{o}F_J^{**}(z, g)}{(\hat{\alpha}^{o}(z))^2} \qquad ((z, g) \in \boldsymbol{a}_C^* \setminus \hat{\alpha}^{-1}(\operatorname{spec}\mathbf{K}_\alpha) \times G).$$

It is clear that Corollaries 1 and 2 remain valid if ${}^{o}F_J^{**}(z, g)$ is replaced by $F_J^{**}(z, g)$.

Finally, the important expression (3.5.4) is valid on the pointwise level. Thus for $(\Lambda, g) \in (\boldsymbol{a}_C^*)^+ \setminus \hat{\alpha}^{-1}(\operatorname{spec}\mathbf{K}) \times G$

$$E(\Lambda, g) = \sum_J \phi(J \mid \Lambda) F_J^*(\Lambda, g).$$

CHAPTER 4

ANALYTIC CONTINUATION

4.1 Introduction

Since we now have

$$E(\Lambda, g) = \sum_{J} \phi(J\,|\,\Lambda)F_J^*(\Lambda, g)\,,$$

where the $F_J^*(\Lambda, g)$ are already holomorphic on $\boldsymbol{a}_C^* \setminus \hat{\alpha}^{-1}(\text{spec } \mathbf{K}_\alpha)$, we turn our attention to the constant term coefficients $\phi(J\,|\,\Lambda)$. A system of linear equations with $\phi(J\,|\,\Lambda)$ as unknowns will be formed on the basis that the Eisenstein series is an eigenfunction for all of $\mathbf{Z}(G)$, that its constant terms have a definite form, and that it is a σ-function. As it happens, because we are dealing with a maximal cuspidal subgroup, the combinatorics of $\phi(J\,|\,\Lambda)$ is exceedingly simple (Lemma 4.3.4). Moreover, it can easily be verified that $\phi(J_1\,|\,\Lambda)$, where $J_1 = (1, 1, n)$, are all constant functions, and therefore they pose no difficulties in analytic continuation. We will solve for the remaining coefficients $\phi(J'\,|\,\Lambda)$ uniquely in terms of these. The key is that, for all appropriately restricted $\Lambda \in (\boldsymbol{a}_C^*)^+ \setminus \hat{\alpha}^{-1}(\text{spec } \mathbf{K}_\alpha)$, nonuniqueness of solution would lead to a nonzero function which has properties similar to the Eisenstein series, except that it is bounded (Lemma 4.3.6). As we shall see (Theorem 4.3.1), this contradicts the self-adjointness of the Casimir operator (Proposition 4.3.1).

To fully establish the meromorphicity of the constant term coefficients $\phi(j, s, n\,|\,\Lambda)$, and of the Eisenstein series itself, we exploit the freedom we have in the choice of the kernel function α. Thus more than one compact operator $\mathbf{K}_\alpha$ will be used. We then arrive at two of the main results of this work, namely, Theorem 4.4.1 and Theorem 4.4.2.

Lastly, we show that the continued Eisenstein series possesses all the characteristic properties of the original. In particular, it is a smooth function of (Λ, g) (Theorem 4.4.3).

4.2. A system of Linear Equations in $\phi(J \mid \Lambda)$

Since now

$$E(\Lambda, g) = F(\Lambda, g) = \sum_J \phi(J \mid \Lambda) F_J^*(\Lambda) = \sum_{j,s,n} \phi(j, s, n \mid \Lambda) F_{j,s,n}^*(\Lambda) \qquad (g \in G)$$

for $\Lambda \in (\mathfrak{a}_C^*)^+ \setminus \hat{\alpha}^{-1}(\text{spec } \mathbf{K}_\alpha)$ where $F_J^*(\Lambda, g) = F_J^{**}(\Lambda, g) + H_J(\Lambda, g)$ are already holomorphic on $\mathfrak{a}_C^* \setminus \hat{\alpha}^{-1}(\text{spec } \mathbf{K}_\alpha)$, the analytic continuation of the Eisenstein series, as we shall see, has largely been reduced to the continuation of the constant term coefficients $\phi(J \mid \Lambda)$.

In continuing $\phi(J \mid \Lambda)$, direct use will be made of the fact that $E(\Lambda, g)$ is an eigenfunction for $\mathbf{Z}(G)$. For $\Lambda \in (\mathfrak{a}_C^*)^+ \setminus \hat{\alpha}^{-1}(\text{spec } \mathbf{K}_\alpha)$, $DE = \chi_\Lambda(D)E$ implies

$$\sum_J \phi(J) D F_J^* = \sum_J \phi(J) \chi_\Lambda(D) F_J^*, \qquad D \in \mathbf{Z}(G), \tag{4.2.1}$$

so we have

$$\sum_J \phi(J \mid \Lambda)[\, D F_J^*(\Lambda, g) - \chi_\Lambda(D) F_J^*(\Lambda, g)\,] = 0 \qquad (g \in G), \quad D \in \mathbf{Z}(G). \tag{4.2.1'}$$

To the equation (4.2.1′) we add three more. Let P_i be a maximal standard cuspidal subgroup associate to P, with $P_i = N_i A_i M_i$. For $\Lambda \in (\mathfrak{a}_C^*)^+ \setminus \hat{\alpha}^{-1}(\text{spec } \mathbf{K}_\alpha)$, $g = n_i a_i m_i k$, where $n_i \in N_i$, $a_i \in A_i$, $m_i \in M_i$, $k \in K$,

$$\int_{N_i \cap \Gamma \setminus N_i} E(\Lambda, ng)\, dn = \sum_{s,n} \phi(i, s, n \mid \Lambda) e^{(-s\Lambda - \rho_i)(\log a_i)} \sigma(k) \Psi_{i,s,n}(m_i)$$

implies

$$\sum_{j,s,n} \phi(j, s, n \mid \Lambda) \int_{N_i \cap \Gamma \setminus N_i} F_{j,s,n}^*(\Lambda, ng)\, dn \tag{4.2.2}$$

$$= \sum_{s,n} \phi(i, s, n \mid \Lambda) e^{(-s\Lambda-\rho_i)(\log a_i)} \sigma(k) \Psi_{i,s,n}(m_i) .$$

Thus

$$(4.2.2') \quad \sum_{j,s,n} \phi(j, s, n \mid \Lambda) \left[\int_{N_i \cap \Gamma \backslash N_i} F^*_{j,s,n}(\Lambda, ng)\, dn - \delta_{ji} e^{(-s\Lambda-\rho_i)(\log a_i)} \sigma(k) \Psi_{i,s,n}(m_i) \right] = 0 \quad (g \in G),$$

i *varies over the same index set as* j (δ_{ji} is the Kronecker delta). On the other hand, if $\{P_l : l \in L\}$ is the set of all rank one standard cuspidal subgroups not associate to P, then

$$\int_{N_l \cap \Gamma \backslash N_l} E(\Lambda, ng)\, dn = 0$$

implies

$$(4.2.3') \qquad \sum_J \phi(J \mid \Lambda) \int_{N_l \cap \Gamma \backslash N_l} F^*_J(\Lambda, ng)\, dn = 0 \qquad (g \in G),\ l \in L .$$

Lastly, for $\Lambda \in (\boldsymbol{a}^*_C)^+ \backslash \hat{\alpha}^{-1}(\text{spec } \mathbf{K}_\alpha)$,

$$E(\Lambda, gk) = \sigma(k) E(\Lambda, g) \qquad (g \in G) \text{ for all } k \in K$$

implies

$$(4.2.4') \qquad \sum_J \phi(J \mid \Lambda)[\, F^*_J(\Lambda, gk) - \sigma(k) F^*_J(\Lambda, g)\,] = 0 \qquad (g \in G),\ k \in K .$$

For simplicity, we shall often use a multiindex J_1 in place of (1, 1, n) (i.e., j = 1, s = 1 = the identity element of $W(\boldsymbol{a}, \boldsymbol{a})$). The coefficients $\phi(J_1 \mid \Lambda)$ are actually constant, so trivially become meromorphic on $\boldsymbol{a}^*_C$.

Proposition 4.2.1. In the equation

$$\int_{N \cap \Gamma \backslash N} E(\Lambda, ng)\, dn = \sum_{s \in W(\boldsymbol{a}, \boldsymbol{a})} e^{(-s\Lambda-\rho)(\log a)} \sigma(k)(M(s \mid \Lambda)\Phi)(m) ,$$

$M(1 \mid \Lambda)$ is the identity transformation on ${}^0L^2(\Gamma_M \backslash M, \sigma_M, \chi)$.

Proof. From Harish-Chandra [2, p. 44] we have

$$\int_{N \cap \Gamma \backslash N} \int_{\Gamma \cap \rho \backslash \Gamma(s)} \Phi_\Lambda(\gamma nam)\, dn = e^{(-s\Lambda-\rho)(\log a)}(M(s \mid \Lambda)\Phi)(m) ,$$

where $\Gamma(s) = \Gamma \cap PyP$, $y \in G_Q$ such that $Ad(y) = s$ on $\boldsymbol{a}$. If $s = 1$, then $\Gamma(s) = \Gamma$, so

$$\begin{aligned} e^{(-\Lambda-\rho)(\log a)}(M(1 \mid \Lambda)\Phi)(m) &= \int_{N \cap \Gamma \backslash N} \Phi_\Lambda(nam)\, dn \\ &= \int_{N \cap \Gamma \backslash N} e^{(-\Lambda-\rho)(\log a)}\Phi(m)\, dn \\ &= e^{(-\Lambda-\rho)(\log a)}\Phi(m) . \end{aligned}$$

Since this is true for all $\Phi \in {}^oL^2(\Gamma_M \backslash M, \sigma_M, \chi)$, $M(1 \mid \Lambda) =$ identity. □

Therefore, we rewrite equations (4.2.1′), (4.2.2′), (4.2.3′), and (4.2.4′):

$$\sum_{J'} \phi(J' \mid \Lambda) A_{D,J'}(\Lambda, g) = - \sum_{J_1} \phi(J_1) A_{D,J_1}(\Lambda, g) , \qquad D \in Z(G) ,$$

$$\sum_{J'} \phi(J' \mid \Lambda) B_{i,J'}(\Lambda, g) = - \sum_{J_1} \phi(J_1) B_{i,J_1}(\Lambda, g) ,$$

i varies over the same index set as j ,

$$\sum_{J'} \phi(J' \mid \Lambda) C_{l,J'}(\Lambda, g) = - \sum_{J_1} \phi(J_1) C_{l,J_1}(\Lambda, g) , \qquad l \in L ,$$

$$\sum_{J'} \phi(J' \mid \Lambda) D_{k,J'}(\Lambda, g) = - \sum_{J_1} \phi(J_1) D_{k,J_1}(\Lambda, g) , \qquad k \in K ;$$

where

$$J' = (j, s', n) ,$$

$$s' \in \begin{cases} W(\boldsymbol{a}, \boldsymbol{a}) \setminus \{1\} & \text{if } j = 1 \\ W(\boldsymbol{a}, \boldsymbol{a}_j) & \text{if } j \neq 1 , \end{cases}$$

$$A_{D,J'}(\Lambda, g) = DF^*_{J'}(\Lambda, g) - \chi_\Lambda(D) F^*_{J'}(\Lambda, g) ,$$

$$B_{i,J'}(\Lambda, g) = B_{i;j,s',n}(\Lambda, g) = \int_{N_i \cap \Gamma \backslash N_i} F^*_{j,s',n}(\Lambda, ng)\, dn$$

$$- \delta_{ji} e^{(-s'\Lambda - \rho_i)(\log a)} \Psi_{i,s',n}(m) ,$$

$$C_{l,J'}(\Lambda, g) = \int_{N_l \cap \Gamma \backslash N_l} F^*_{J'}(\Lambda, ng)\, dn ,$$

$$D_{k,J'}(\Lambda, g) = F^*_{J'}(\Lambda, gk) - \sigma(k) F^*_{J'}(\Lambda, g) .$$

For each $g \in G$, these functions are all holomorphic on $\boldsymbol{a}^*_{\mathbb{C}} \backslash \hat{\alpha}^{-1}(\text{spec } \mathbf{K}_\alpha)$, and have at worst poles at $\hat{\alpha}^{-1}(\text{spec } \mathbf{K}_\alpha \backslash \{0\})$.

If we fix $\Lambda \in (\boldsymbol{a}^*_{\mathbb{C}})^+ \backslash \hat{\alpha}^{-1}(\text{spec } \mathbf{K}_\alpha)$, and let g vary over G, we obtain a system of linear equations with $A_{D,J'}(g)$, $B_{i,J'}(g)$, $C_{l,J'}(g)$, $D_{k,J'}(g)$ as coefficients and $\phi(J')$ as unknowns:

$$\sum_{J'} \phi(J') A_{D,J'}(g) = - \sum_{J_1} \phi(J_1) A_{D,J_1}(g) , \qquad D \in \mathbf{Z}(g) ,$$

$$\sum_{J'} \phi(J') B_{i,J'}(g) = - \sum_{J_1} \phi(J_1) B_{i,J_1}(g), \qquad i \textit{ varies over the same index set as } j ,$$

$$\sum_{J'} \phi(J') C_{l,J'}(g) = - \sum_{J_1} \phi(J_1) C_{l,J_1}(g), \qquad l \in L ,$$

$$\sum_{J'} \phi(J') D_{k,J'}(g) = - \sum_{J_1} \phi(J_1) D_{k,J_1}(g) , \qquad k \in K ;$$

$$g \in G .$$

From this system of linear equations in $\phi(J')$, *which is known to be consistent,* we shall be able to express these constant term coefficients in terms of $A_{D,J'}(g)$, $B_{i,J'}(g)$, $C_{l,J'}(g)$, and $D_{k,J'}(g)$ $(g \in G)$.

Denote the cardinality of $\{\phi(J')\}$ by m. The m-tuples $\{A_{D,J'}\}_{J'}$, $\{B_{i,J'}\}_{J'}$, $\{C_{l,J'}\}_{J'}$, $\{D_{k,J'}\}_{J'}$ correspond to vectors in $\mathbf{C}^m$. If the dimension of the span of these vectors is equal to m, then there exist m equations from the system whose determinant is nonzero, and one can solve for

$\{\phi(J')\}$ immediately. Suppose the contrary, i.e., the number of independent equations is less than m. Then one gets more than one solution. Let $\{X_{J'}^{(1)}\}$ and $\{X_{J'}^{(2)}\}$ be two distinct solutions, and set $\phi^-(J') = X_{J'}^{(1)} - X_{J'}^{(2)}$. The complex numbers $\phi^-(J')$ are not all zero. Set $\phi^-(J_1) = 0$. We claim that the function $F^-(g) = \sum_J \phi^-(J)F_J^*(g)$, which satisfies equations (4.2.1′), (4.2.2′), (4.2.3′), and (4.2.4′) also, is nonzero. By the equation (4.2.2), if $F^- = 0$, then

$$\sum_{s,n} \phi^-(i, s, n)e^{(-s\Lambda-\rho_i)(\log a_i)}\sigma(k)\Psi_{i,s,n}(m_i) \equiv 0 \qquad \text{for each } i\,.$$

However, we have the following lemma. Note that $\Lambda \in (\boldsymbol{a}_C^*)^+$ implies $\Lambda \neq 0$.

Lemma 4.2.1. Let P_i be a maximal standard cuspidal subgroup associate to P, $P_i = N_iA_iM_i$. For every nonzero $\Lambda \in \boldsymbol{a}_C^*$, the functions on G:

$$e^{(-s\Lambda-\rho_i)(\log a_i)}\sigma(k)\Psi_{i,s,n}(m_i)$$

($g = n_ia_im_ik$, where $n_i \in N_i$, $a_i \in A_i$, $m_i \in M_i$, $k \in K$), where s varies over $W(\boldsymbol{a}, \boldsymbol{a}_i)$ and n as in $\{\Psi_{i,s,n}\}$, a basis for ${}^0L^2(\Gamma_{M_i} \backslash M_i, \sigma_{M_i}, {}^s\chi)$, are linearly independent.

Proof. Suppose

$$\sum_{s,n} \alpha_{s,n}e^{(-s\Lambda-\rho_i)(\log a_i)}\sigma(k)\Psi_{i,s,n}(m_i) \equiv 0$$

for some $\alpha_{s,n} \in \mathbf{C}$. Since $\sigma(k)$ never vanishes,

$$\sum_{s,n} \alpha_{s,n}e^{(-s\Lambda-\rho_i)(\log a_i)}\Psi_{i,s,n}(m_i) \equiv 0\,.$$

Now

$$\sum_{s,n} \alpha_{s,n}e^{(-s\Lambda-\rho_i)(\log a_i)}\Psi_{i,s,n}(m_i) = \sum_s e^{(-s\Lambda-\rho_i)(\log a_i)} \sum_n \alpha_{s,n}\Psi_{i,s,n}(m_i)\,.$$

If v is a nonzero element of $\boldsymbol{a}_i \subset (\boldsymbol{a}_i)_C$, then $\log a_i = xv$ with $x \in \mathbf{R}$. Write $(-s\Lambda-\rho_i)(\log a_i) = \beta_s x$ with $\beta_s \in \mathbf{C}$. Since $\Lambda \neq 0$, the β_s ($s \in W(\boldsymbol{a}, \boldsymbol{a}_i)$) are all distinct. We now have

$$\sum_s e^{\beta_s x}\left[\sum_n \alpha_{s,n}\Psi_{i,s,n}(m_i)\right] \equiv 0 \qquad \text{for } x \in \mathbf{R},\ m_i \in M_i\,.$$

But the functions $e^{\beta_s x}$ ($s \in W(\boldsymbol{a}, \boldsymbol{a}_i)$) are linearly independent, because, for instance, the value of the Wronskian at $x = 0$ is

$$\det\begin{bmatrix} 1 & 1 & \cdots & 1 \\ \beta_1 & \beta_2 & \cdots & \beta_N \\ \vdots & & & \vdots \\ \beta_1^{N-1} & \beta_2^{N-1} & \cdots & \beta_N^{N-1} \end{bmatrix}, \qquad (W(\boldsymbol{a}, \boldsymbol{a}_1) = \{s_1, \ldots, s_N\},\ \beta_1 = \beta_{s_1}, \ldots, \beta_N = \beta_{s_N}).$$

Then

$$\sum_s e^{\beta_s x}\left[\sum_n \alpha_{s,n}\Psi_{i,s,n}(m_i)\right] \equiv 0$$

implies

$$\sum_n \alpha_{s,n}\Psi_{i,s,n}(m_i) \equiv 0 \qquad \text{for } m_i \in M_i\,.$$

Since $\{\Psi_{i,s,n}\}$ is a basis of ${}^oL^2(\Gamma_{M_i} \backslash M_i, \sigma_{M_i}, {}^s\chi)$, we can conclude that $\alpha_{s,n} = 0$ for all n, and this is true for each $a \in W(\boldsymbol{a}, \boldsymbol{a}_i)$. □

4.3. Uniqueness of Solution

In the last section we saw that if the determinant of every m equations from the linear system vanished, then we could construct a *nonzero* function $F^-(g)$ which possessed properties similar to those of $F(g)$, or the Eisenstein series; the crucial difference between $F(g)$ and $F^-(g)$ is that $\phi^-(J_1) = 0$ for the latter. However, as we shall see later, for an appropriately restricted $\Lambda \in (\boldsymbol{a}_C^*)^+ \backslash \hat{\alpha}^{-1}(\text{spec } \mathbf{K}_\alpha)$, this cannot happen. Consequently, with such a choice of Λ, $\{\phi(J)\}$ can be expressed in terms of $A_{D,J}(\Lambda, g)$, $B_{i,j}(\Lambda, g)$, $C_{l,J}(\Lambda, g)$, $D_{k,J}(\Lambda, g)$ $\quad (g \in G)$. First we list and develop the main properties of $F^-(g)$ in a number of lemmas. Recall that

$$F^-(g) = \sum_J \phi^-(J) F_J^*(g) .$$

It is clear that $F^-(g)$ is Γ-automorphic, since each of the $F_J^*(g)$ is. Also, because of the equation (4.2.4′), $F^-(g)$ is a σ-function.

Lemma 4.3.1. The function $F^-(g)$ is slowly increasing.

Proof. By Corollary 2 to Theorem 3.6.1, $F_J^{**}(g)$ is bounded, hence trivially slowly increasing. From Section 3.4, we know that $H_J(g)$ is slowly increasing. So the functions $F_J^*(g)$ are all slowly increasing, and therefore $F^-(g)$ too. □

Since $\{\phi^-(J)\}$ satisfies the equation (4.2.1), we have:

Lemma 4.3.2. The function $F^-(g)$ is an eigenfunction for all of $\mathbf{Z}(G)$ with the same set of eigenvalues as the Eisenstein series.

Also, because of the equation (4.2.3′), we have:

Lemma 4.3.3. The constant term of $F^-(g)$ with respect to any maximal standard cuspidal subgroup not associate to P is zero.

On the other hand, if P_i is a maximal standard cuspidal subgroup associate to P, then the equation (4.2.2) implies that the constant term of $F^-(g)$ with respect to P_i is

$$\sum_{s,n} \phi^-(i, s, n) e^{(-s\Lambda - \rho_i)(\log a_i)} \sigma(k) \Psi_{i,s,n}(m_i) ,$$

where $s \in W(\boldsymbol{a}, \boldsymbol{a}_i)$. The next lemma greatly clarifies the combinatorial structure of these constant terms. It is very simple.

Lemma 4.3.4. *The number of maximal standard cuspidal subgroups associate to* P_1, *a given maximal standard cuspidal subgroup, is either* 1 *or* 2. *In the first case,*

$$W(\mathbf{a}_1, \mathbf{a}_1) = \{1, s\}, \quad \text{where } sH_1 = -H_1 \text{ for all } H_1 \in \mathbf{a}_1 .$$

In the second case, suppose P_2 *is the other maximal standard cuspidal subgroup, if* α_1, *respectively* α_2, *is the simple root of* (P_1, A_1), *respectively* (P_2, A_2), *and* $H_1 \in \mathbf{a}_1$, $H_2 \in \mathbf{a}_2$ *such that* $\alpha_1(H_1) = 1$, $\alpha_2(H_2) = 1$, *then* $W(\mathbf{a}_1, \mathbf{a}_2) = \{s\}$, *where* $sH_1 = -H_2$.

(See Harish-Chandra [2, p. 89], and Langlands [1, p. 126].)

We remind ourselves that by virtue of the construction of $F^-(g)$, $\phi^-(J_1) = \phi^-(1, 1, n) = 0$. Consequently, in the first case, the only nonzero constant term of $F^-(g)$ is the one with respect to $P = P_1$, it equals

$$\sum_n \phi^-(1, s, n) e^{(-s\Lambda-\rho_1)(\log a_1)} \sigma(k) \Psi_{1,s,n}(m_1) ;$$

in the second case, since $W(\mathbf{a}, \mathbf{a}) = \{1\}$ and $\phi^-(1, 1, n) = 0$, the only nonzero constant term of $F^-(g)$ is the one with respect to P_2, which is

$$\sum_n \phi^-(2, s, n) e^{(-s\Lambda-\rho_2)(\log a_2)} \sigma(k) \Psi_{2,s,n}(m_2) .$$

Hereafter, the index i will either be 1 or 2, $H \in \mathbf{a}$ such that $\alpha(H) = 1$.

It is the boundedness of $F^-(g)$ that really sets it apart from $F(g)$. Before we can demonstrate the boundedness of $F^-(g)$ we need one more fact from structure theory. The references we cited for Lemma 3.4.1 are serviceable here. The Levi component M_r of a maximal standard cuspidal subgroup P_r has the Iwasawa decomposition $M_r = N_r' A_r' K_r'$, where $N_r' \subset N_0$, $A_r' \subset A_0$, $K_r' \subset K$. Denote the Lie algebra of A_r' by $\mathbf{a}_r'$. Let $\langle\ ,\ \rangle$ be the Killing form on $\mathbf{g} \times \mathbf{g}$. For each α_p in the set of simple roots $\{\alpha_p : p = 1, \ldots, m\}$ of $(\mathbf{p}_0, \mathbf{a}_0)$, let Y_{α_p} be the element of $\mathbf{a}_0$ defined by $\alpha_p(Y) = \langle Y, Y_{\alpha_p} \rangle$ for all $Y \in \mathbf{a}_0$. Also, let $\{Y_p\}$ be the basis of $\mathbf{a}$ dual to $\{\alpha_p\}$. Then

$$\boldsymbol{a}_r' = \{X \in \boldsymbol{a}_0 : \langle X, Y \rangle = 0 \quad \forall\, Y \in \boldsymbol{a}_r\}$$

$$= \sum_{p \neq r} \mathbf{R} Y_{\alpha_p} .$$

Lemma 4.3.5. For each $r \in \{1, \dots, m\}$, there exist $C_{rp} \geq 0$ $(p = 1, \dots, m)$ with $C_{rr} > 0$ such that $\sum_p C_{rp}\alpha_p(Y) = 0$ for all $Y \in \boldsymbol{a}_r'$.

Proof. Let $[\langle \alpha_p, \alpha_q \rangle]_{1 \leq p,\, q \leq m}$ be the matrix defined by

$$\langle \alpha_p, \alpha_q \rangle = \langle Y_{\alpha_p}, Y_{\alpha_q} \rangle = \alpha_q(Y_{\alpha_p}) .$$

It is known that the inverse $[\langle \alpha_p, \alpha_q \rangle]^{-1}$ of this matrix has positive entries (Harish-Chandra [2, p. 24], Langlands [1, p. 12]). We simply let $[\langle \alpha_p, \alpha_q \rangle]^{-1}$ be $[C_{rp}]$. Now $\langle Y_r , Y_{\alpha_p} \rangle = \alpha_p(Y_r) = 0$ for all $p \neq r$ implies $\langle Y_r , Y \rangle = 0$ for all $Y \in \boldsymbol{a}_r'$. On the other hand, for every $q \in \{1, \dots , n\}$, we have

$$\alpha_p(\sum_p C_{rp} Y_{\alpha_p}) = \sum_p C_{rp}\alpha_q(Y_{\alpha_p})$$

$$= \sum_p C_{rp}\langle \alpha_p , \alpha_q \rangle$$

$$= \delta_{rq} ,$$

which implies $\sum_p C_{rp} Y_{\alpha_p} = Y_r$. Then

$$\sum_p C_{rp}\alpha_p(Y) = \langle Y, \sum_p C_{rp} Y_{\alpha_p} \rangle = \langle Y, Y_r \rangle = 0$$

for all $Y \in \boldsymbol{a}_r'$, and the lemma is proved. □

Now we can deduce the crucial assertion about $F^-(g)$.

Lemma 4.3.6. *If* $\Lambda \in (\mathfrak{a}_C^*)^+ \setminus \hat{\alpha}^{-1}(\text{spec } \mathbf{K}_\alpha)$ *is such that* $[(\mathrm{Re}\Lambda)(H) - \rho(H)] + [\rho(H) - \rho_i(H_i)] > 0$, *then* $F^-(g)$ *is bounded.*

Proof. Let S_0 be the Siegel domain, and F be the fundamental domain of Chapters 2 and 3. Since $F^{-1}(g)$ is Γ-automorphic, slowly increasing, a σ-function, and an eigenfunction for all of $\mathbf{Z}(G)$,

$$F^- - \sum_r [\, \delta_r F^{-r} \,]_F$$

is bounded on G by Theorem 3.5.2. Of course, in the first case,

$$F^{-r}(g) = \sum_n \phi^-(1, s, n) e^{(-s\Lambda-\rho_1)(\log a_1)} \sigma(k) \Psi_{1,s,n}(m_1) \qquad \text{for } r = 1 ,$$

$$= 0 \qquad \text{for } r \neq 1 ;$$

in the second case,

$$F^{-r}(g) = \sum_n \phi^-(2, s, n) e^{(-s\Lambda-\rho_2)(\log a_2)} \sigma(k) \Psi_{2,s,n}(m_2) \qquad \text{for } r = 2 ,$$

$$= 0 \qquad \text{for } r \neq 2 .$$

Fix $H_i \in \mathfrak{a}_i$ such that $\alpha_i(H_i) = 1$, where α_i is the simple root of $(\mathfrak{p}_i, \mathfrak{a}_i)$. For $g = n_i a_i m_i k$, where $n_i \in N_i$, $a_i \in A_i$, $m_i \in M_i$, $k \in K$,

$$\alpha_i(H_i(g)) = \alpha_i(H_0(g)) - \alpha_i(H_0(m_i)) .$$

Now, by Lemma 4.3.5,

$$\alpha_i(H_0(m_i)) = \sum_t C_{it} \alpha_t(H_0(m_i))$$

$$= \sum_t C_{it} \alpha_t(H_0(g)) ,$$

where $C_{it} \leq 0$, $t = 1, \ldots, i-1, i+1, \ldots, m$ $(m = \dim \mathfrak{a}_0)$. For $g \in S_0$, we then have

$$\alpha_i(H_i(g)) \leq t_0(1 - \sum_t C_{it}) .$$

Therefore, if $H_i(g) = xH_i$, $x \in \mathbf{R}$, $\alpha_i(H_i(g)) = x\alpha_i(H_i) = x$, and

$$-\infty \leq x \leq t_0(1 - \sum_t C_{it}) .$$

Recall that $(\boldsymbol{a}_C^*)^+$, the domain of definition of the Eisenstein series, consists of all $\Lambda \in \boldsymbol{a}_C^*$ such that

$$(\mathrm{Re}\Lambda)(H) - \rho(H) > 0 .$$

In

$$\sum_n \phi^-(i, s, n) e^{(-s\Lambda-\rho_i)(\log a_i)} \Psi_{i,s,n}(m_i) ,$$

where $s \in W(\boldsymbol{a}, \boldsymbol{a}_i) \setminus \{1\}$,

$$e^{(-s\Lambda-\rho_i)(\log a_i)} = e^{(-s\Lambda-\rho_i)(H_i(g))} = e^{(-s\Lambda-\rho_i)(xH_i)} = e^{x(-s\Lambda)(H_i)-x\rho_i(H_i)} ,$$

by Lemma 4.3.4,

$$= e^{x\Lambda(H)-x\rho_i(H_i)}$$

$$= e^{x\{ [\Lambda(H)-\rho(H)]+[\rho(H)-\rho_i(H_i)] \}} .$$

Hence

$$| e^{(-s\Lambda-\rho_i)(\log a_i)} | = e^{x\{ [\mathrm{Re}\Lambda)(H)-\rho(H)]+[\rho(H)-\rho_i(H_i)] \}} .$$

If $\Lambda \in \boldsymbol{a}_C^*$ is such that $[(\mathrm{Re}\Lambda)(H) - \rho(H)] + [\rho(H) - \rho_i(H_i)] > 0$, then $e^{(-s\Lambda-\rho_i)(\log a_i)}$ is bounded on $S_0 \supset F$. Also, being cusp forms, all the functions $\Psi_{i,s,n}(m_i)$ are bounded. Of course, $\sigma(k)$ is bounded. Thus

$$F^- = (F^- - [\, \delta_i F^{-\,i}]_F) + [\, \delta_i F^{-\,i}]_F ,$$

which is the sum of two bounded functions on G. □

We will next prove a theorem which enables us to deduce uniqueness of solution of the system of linear equations in the last section for appropriately restricted Λ. The theorem involves a distinguished element of **Z**(G), namely, the Casimir operator ω. As an element of U($\boldsymbol{g}$), the *Casimir operator* ω can be given as follows (Knapp [1, pp. 209-211]): Recall the *Cartan decomposition* $\boldsymbol{g} = \boldsymbol{k} \oplus \boldsymbol{p}$. Let $\{X_i\}$ be an orthonormal basis of $\boldsymbol{k}$ and $\{Y_j\}$ be an orthonormal basis of $\boldsymbol{p}$, both with respect to $\langle\ ,\ \rangle_\theta$, where $\langle X, Y \rangle_\theta = -\langle X, \theta Y \rangle = -\mathrm{Tr}(\mathrm{ad}X\mathrm{ad}\theta Y)$ for $X, Y \in \boldsymbol{g}$, θ is the *Cartan involution* on $\boldsymbol{g}$. Then

$$\omega = -\sum_i X_i^2 + \sum_j Y_j^2 .$$

Proposition 4.3.1 indicates the self-adjointness of ω. If p and p are smooth Γ-automorphic functions on G, define

$$[p, q] = \omega p \cdot \overline{q} - p \cdot \overline{\omega q} .$$

<u>Proposition 4.3.1.</u> Suppose p and q are smooth Γ-automorphic functions on G. If one of them is *compactly supported on* $\Gamma \backslash G$, then

$$\int_{\Gamma \backslash G} [p, q]\, dg = 0 .$$

<u>Proof</u>. Since $\omega = -\sum_i X_1^2 + \sum_j Y_j^2$, where $X_1, Y_j \in \boldsymbol{g}$, the result will follow if

$$\int_{\Gamma \backslash G} X^2 p \cdot \overline{q} - p \cdot \overline{X^2 q}\, dg = 0 \qquad \text{for all } X \in \boldsymbol{g} .$$

Let us assume without loss of generality that p is compactly supported on $\Gamma \backslash G$. In turn, we need only show that

$$\int_{\Gamma \backslash G} Xp \cdot q\, dg = -\int_{\Gamma \backslash G} p \cdot \overline{Xq}\, dg \qquad \text{for all } X \in \boldsymbol{g} ,$$

for then

$$\int_{\Gamma \backslash G} X^2 p \cdot \overline{q}\, dg = -\int_{\Gamma \backslash G} Xp \cdot \overline{Xq}\, dg = \int_{\Gamma \backslash G} p \cdot X^2 q\, dg .$$

By definition,

$$\int_{\Gamma\backslash G} Xp(g)\overline{q}(g)\,dg = \int_{\Gamma\backslash G} \left[\frac{d}{dt}\,p(g\exp tX)\right]_{t=0} \overline{q}(g)\,dg$$

$$= \int_{\Gamma\backslash G} \lim_{t\to 0}\,(p(g\exp tX) - p(g))/t\;\overline{q}(g)\,dg$$

$$= \int_{\Gamma\backslash G} \lim_{t\to 0}\,(Xp)(g\exp t_0X)\cdot\overline{q}(g)\,dg\,,$$

where $(p(g\exp tX) - p(g))/t = (Xp)(g\exp t_0X)$, $|t_0| \leq |t| \leq \varepsilon$. Like p, Xp too is compactly supported on $\Gamma\backslash G$. Denote by π the projection map from G onto $\Gamma\backslash G$. If $\Delta \subset \Gamma\backslash G$ is the compact support of Xp on $\Gamma\backslash G$, then $\pi^{-1}(\Delta) = \Gamma\Omega$ for some compact subset Ω of G. It is clear that $(Xp)(g\exp t_0X) \neq 0$ implies $g \in \pi(\Gamma\Omega\exp - t_0X) = \pi(\Omega\exp - t_0X)$. Therefore, if ε is sufficiently small, there is a compact subset Δ^* of $\Gamma\backslash G$ such that

$$\int_{\Gamma\backslash G} Xp(g)\overline{q}(g)\,dg = \int_{\Delta^*} \lim_{t\to 0}\,(Xp)(g\exp t_0X)\cdot\overline{q}(g)\,dg\,,$$

and

$$\lim_{t\to 0} \int_{\Gamma\backslash G} (Xp)(g\exp t_0X)\cdot\overline{q}(g)\,dg = \lim_{t\to 0} \int_{\Delta^*} (Xp)(g\exp t_0X)\cdot\overline{q}(g)\,dg\,.$$

Since $\overline{q}(g)$ is continuous on $\Gamma\backslash G$, and so bounded over Δ^*, we can apply the dominated convergence theorem to deduce that

$$\int_{\Gamma\backslash G} Xp(g)\overline{q}(g)\,dg = \lim_{t\to 0} \int_{\Gamma\backslash G} (Xp)(g\exp t_0X)\cdot\overline{q}(g)\,dg$$

$$= \lim_{t\to 0} \int_{\Gamma\backslash G} (p(g\exp tX) - p(g))/t\;\overline{q}(g)\,dg$$

$$= \left[\frac{d}{dt} \int_{\Gamma\backslash G} p(g\exp tX)\overline{q}(g)\,dg\right]_{t=0}$$

$$= \left[\frac{d}{dt} \int_{\Gamma\backslash G} p(g)\overline{q}(g\exp - tX)\,dg\right]_{t=0}\,.$$

The change of variable for the last integral is justified by the fact that the map

$\Gamma\backslash G \to \Gamma\backslash G : g \mapsto g \exp -tX$ is measure preserving. This integral can likewise be shown to be equal to

$$\int_{\Gamma\backslash G} p(g)\left[\frac{d}{dt}\,\overline{q}(g\exp -tX)\right]_{t=0} dg$$

$$= \int_{\Gamma\backslash G} p(g)\cdot -X\overline{q}(g)\,dg = -\int_{\Gamma\backslash G} p(g)X\overline{q}(g)\,dg\,.$$

If $q(g) = u(g) + iv(g)$,

$$X\overline{q}(g) = \left[\frac{d}{dt}\,u(g\exp tX)\right]_{t=0} - i\left[\frac{d}{dt}\,v(g\exp tX)\right]_{t=0}$$

$$= (Xu)(g) - i(Xv)(g)$$

$$= \overline{Xq}(g)\,.$$

This completes the proof. □

<u>Theorem 4.3.1.</u> *For a bounded Γ-automorphic eigenfunction of the Casimir operator which is both* K*-finite and* $\mathbf{Z}(G)$*-finite, the corresponding eigenvalue of* ω *is real.*

<u>Proof.</u> Let f be such a function. Since f is K-finite and $\mathbf{Z}(G)$-finite, Theorem 2.2.2 guarantees the existence of a smooth compactly supported function τ such that $\tau * f = f$. Then

$$f(g) = \tau * f\,(g) = \int_G \tau(h^{-1}g)f(h)\,dh = \int_G \tau(h)f(gh^{-1})\,dh$$

$$= \int_\Omega \tau(h)f(gh^{-1})\,dh\,,$$

where Ω is the compact support of α. Let $\{C_n\colon n = 1, 2, \cdots\}$ be a sequence of compact subsets of G such that $C_n \subset C_{n+1}$ and $\bigcup_n C_n = G$. Denote by π the projection map from G onto $\Gamma\backslash G$. Let F be the fundamental domain of Chapters 2 and 3. For each n, let β_n be $[\,\chi_n\,]_F$, where χ_n is the characteristic function of C_n. Clearly, $|\beta_n f\,(z)| \le |f(z)|$ and $\beta_n f\,(z) \to f(z)$ for all $z \in G$.

Also, $\beta_n f \in L^2(\Gamma \backslash G)$. Define $f_n = \tau * (\beta_n f)$. For all $g \in G$,

$$f_n(g) = \int_\Omega \tau(h)\beta_n f(gh^{-1})\,dh$$

$$\rightarrow \int_\Omega \tau(h) f(gh^{-1})\,dh = f(g)$$

by the dominated convergence theorem. Since $\beta_n f(gh^{-1}) \neq 0$ implies $gh^{-1} \in \Gamma C_n$, for $\tau(h)\beta_n f(gh^{-1}) \neq 0$, $g \in \Gamma C_n \Omega$. Thus $f_n(g)$ are all compactly supported on $\Gamma \backslash G$. Now, because f_n are smooth Γ-automorphic functions compactly supported on $\Gamma \backslash G$, Proposition 4.3.1 asserts

$$0 = \int_{\Gamma \backslash G} [f_n, f]\,dg = \int_{\Gamma \backslash G} \omega f_n \cdot \overline{f}\,dg - \int_{\Gamma \backslash G} f_n \cdot \overline{\omega f}\,dg. \tag{4.3.1}$$

We want to show that the equation (4.3.1) holds in the limit as $n \rightarrow \infty$. Referring to the remark after Proposition 2.3.4, we have

$$\omega f_n = \omega(\tau * (\beta_n f)) = (\omega\tau) * (\beta_n f),$$

where $\omega\tau$ too is smooth and compactly supported, so likewise

$$\omega f_n(g) \rightarrow \omega\tau * f(g) = \omega(\tau * f)(g) = \omega f(g) \qquad \text{for all } g \in G.$$

By hypothesis f is bounded. And since f is an eigenfunction of ω, ωf too is bounded. From

$$|f_n(g)| \leq \int_\Omega |\tau(h)|\,|\beta_n f(gh^{-1})|\,dh \leq \int_\Omega |\tau(h)|\,|f(gh^{-1})|\,dh \quad \text{for } g \in G,$$

we see that the boundedness of f implies that the functions f_n are uniformly bounded. A similar argument shows that the functions ωf_n are uniformly bounded also. As $\mathrm{Vol}(\Gamma \backslash G) < \infty$, the dominated convergence theorem allows us to pass the equation (4.3.1) to limit, and obtain:

$$0 = \int_{\Gamma \backslash G} [f, f]\,dg = \int_{\Gamma \backslash G} \omega f \cdot \overline{f}\,dg - \int_{\Gamma \backslash G} f \cdot \overline{\omega f}\,dg.$$

If $\omega f = \lambda f$ for some $\lambda \in \mathbf{C}$, then

$$0 = (\lambda - \overline{\lambda}) \int_{\Gamma \backslash G} f \cdot \overline{f}\,dg.$$

So f $\neq$ 0 implies $\lambda = \bar{\lambda}$, i.e., λ is real. □

Recall that for a fixed $\Lambda \in (\boldsymbol{a}_C^*)^+ \setminus \hat{\alpha}^{-1}(\text{spec } \mathbf{K}_\alpha)$, $F^-(g)$ yields the same eigenvalue of ω as $E(\Lambda, g)$, or, as

$$\Phi_\Lambda(g) = \sigma(k)\Phi(m)e^{(-\Lambda-\rho)(\log a)} .$$

This is denoted by $\chi_\Lambda(\omega)$. Therefore, by Theorem 4.3.1, if $\Lambda \in (\boldsymbol{a}_C^*)^+ \setminus \hat{\alpha}^{-1}(\text{spec } \mathbf{K}_\alpha)$ is such that $[\text{Re}(\Lambda)(H) - \rho(H)] + [\rho(H) - \rho_i(H_i)] > 0$ and $\chi_\Lambda(\omega)$ is nonreal, the existence of a nonzero $F^-(g)$ leads to a contradiction.

Lemma 4.3.7. For $\Phi \in {}^0L^2(\Gamma_M \setminus M, \sigma_M, \chi)$, and $H \in \boldsymbol{a}$ such that $\langle H, H \rangle_\theta = 1$,

$$\chi_\Lambda(\omega) = \chi(\omega_m) + \Lambda(H)^2 - \rho(H)^2 \qquad \text{for} \Lambda \in \boldsymbol{a}_C^* ,$$

where $\omega_m \in \mathbf{Z}(M)$. Thus

$$\omega\Phi_\Lambda = [\chi(\omega_m) + \Lambda(H)^2 - \rho(H)^2]\Phi_\Lambda \qquad \text{for} \Lambda \in \boldsymbol{a}_C^* .$$

(See Harish-Chandra [2, pp. 88, 29].)

Clearly, $\chi_\Lambda(\omega)$ is a nonconstant entire function of Λ. There certainly exists $\Lambda \in (\boldsymbol{a}_C^*)^+ \setminus \hat{\alpha}^{-1}(\text{spec } \mathbf{K}_\alpha)$ with $[\text{Re}(\Lambda)(H) - \rho(H)] + [\rho(H) - \rho_i(H_i)] > 0$, such that $\chi_\Lambda(\omega)$ is nonreal. We know that for any such choice of Λ there are m equations from the system whose determinant is nonzero. For $\Lambda \in (\boldsymbol{a}_C^*)^+ \setminus \hat{\alpha}^{-1}(\text{spec } \mathbf{K}_\alpha)$, let $D(\Lambda)$ denote the determinant of these m equations, which depends on Λ. We have thus established the key result of this chapter:

Theorem 4.3.2. There exists an open set $U \subset (\boldsymbol{a}_C^*)^+ \setminus \hat{\alpha}^{-1}(\text{spec } \mathbf{K}_\alpha)$, and holomorphic functions $D(\Lambda)$, $D_{J'}(\Lambda)$ on $\boldsymbol{a}_C^* \setminus \hat{\alpha}^{-1}(\text{spec } \mathbf{K}_\alpha)$, which have either removable singularities or poles at the points of $\hat{\alpha}^{-1}(\text{spec}\mathbf{K}_\alpha \setminus \{0\})$, such that

$$\phi(J' \mid \Lambda) = \frac{D_{J'}(\Lambda)}{D(\Lambda)} \qquad \text{for } \Lambda \in U .$$

4.4. Meromorphicity of $\phi(J \mid \Lambda)$ and the Eisenstein Series

The conclusion of Section 4.3 is that there exists an open set $U \subset (\boldsymbol{a}_{\mathbf{C}}^*)^+ \setminus \hat{\alpha}^{-1}(\mathrm{spec}\ \mathbf{K}_\alpha)$ and holomorphic functions D, $D_{J'}$ on $\boldsymbol{a}_{\mathbf{C}}^* \setminus \hat{\alpha}^{-1}(\mathrm{spec}\ \mathbf{K}_\alpha)$, which have either removable singularities or poles at the points of $\hat{\alpha}^{-1}(\mathrm{spec}\mathbf{K}_\alpha \setminus \{0\})$, such that $\phi(J' \mid \Lambda) = D_{J'}(\Lambda)/ D(\Lambda)$ for $\Lambda \in U$. Denote by D the set of zeros of D in $\boldsymbol{a}_{\mathbf{C}}^* \setminus \hat{\alpha}^{-1}(\mathrm{spec}\ \mathbf{K}_\alpha)$. Each $\phi(J')$, respectively $D_{J'} / D$, is defined and holomorphic on $(\boldsymbol{a}_{\mathbf{C}}^*)^+$, respectively $\boldsymbol{a}_{\mathbf{C}}^* \setminus \hat{\alpha}^{-1}(\mathrm{spec}\ K_\alpha) \cup D$. We will show that $\hat{\alpha}^{-1}(\mathrm{spec}\ \mathbf{K}_\alpha) \cup D$ is a countable closed set and that $\phi(J')$ can be extended to meromorphic functions on $\boldsymbol{a}_{\mathbf{C}}^*$ which have either removable singularities or poles at the points of $\hat{\alpha}^{-1}(\mathrm{spec}\ \mathbf{K}_\alpha) \cup D$; eventually the same will be said of the Eisenstein series itself.

We begin with an elementary topological fact about the complex plane.

Proposition 4.4.1. Let F be a subset of the complex plane **C** with a discrete set of limit points. The set $\mathbf{C} \setminus F$ is connected.

Proof. It suffices to prove that $\mathbf{C} \setminus F$ is path connected. Given $x, y \in \mathbf{C} \setminus F$, we want a continuous mapping $\beta : [0, 1] \to \mathbf{C} \setminus F$ such that $\beta(0) = x$, $\beta(1) = y$. Connect x and y by a continuous path β_0 in **C**. The set F' has no limit points, so $\{\beta_0(t) : t \in [0, 1]\}$, being a compact subset of **C**, can have only finitely many points of F'; and because each point of F' is isolated, we can modify β_0 to a continuous path β_1 in **C** connecting x and y which avoids F' completely. Now by construction $\{\beta_1(t) : t \in [0, 1]\}$ can have only finitely many points in F, and all these are isolated. So β_1 can be modified to a continuous path β in **C** connecting x and y which avoids F completely. It is clear that $\beta: [0, 1] \to \mathbf{C} \setminus F$ is continuous. □

Recall from Section 3.6 that $\hat{\alpha}^{-1}(\mathrm{spec}\ \mathbf{K}_\alpha)$ is a countable closed set with a discrete set of limit points which is either empty or equal to $\hat{\alpha}^{-1}(\{0\})$. So we have:

Corollary 1. The set $a_C^* \setminus \hat{\alpha}^{-1}(\text{spec } \mathbf{K}_\alpha)$ is connected.

Now $D(\Lambda)$ is not identically zero, it follows that D is a countable set whose set of limit points D' is contained in $\hat{\alpha}^{-1}(\text{spec } \mathbf{K}_\alpha)$. If $\Lambda_0 \in \hat{\alpha}^{-1}(\text{spec } \mathbf{K}_\alpha \setminus \{0\})$, $\Lambda_0 \notin D'$. For $D(\Lambda)$ has at worst a pole at Λ_0, were $\Lambda_0 \in D'$, then $D(\Lambda)$ would have to be identically zero in a neighborhood of Λ_0, which would imply that $D(\Lambda)$ is identically zero on an open subset of $a_C^* \setminus \hat{\alpha}^{-1}(\text{spec } \mathbf{K}_\alpha)$, this leads to a contradiction because of Corollary 1. So $D' \subset \hat{\alpha}^{-1}(\{0\})$. We then have:

Corollary 2. The set $(a_C^*)^+ \setminus \hat{\alpha}^{-1}(\text{spec } \mathbf{K}_\alpha) \cup D$ is connected.

Proof. In Proposition 4.4.1, **C** can be replaced by any path connected open subset of it, in particular $(a_C^*)^+$ (a_C^* identified with **C**). Then we have

$$[\,\hat{\alpha}^{-1}(\text{spec } \mathbf{K}_\alpha) \cup D\,]'' = (\,[\,\hat{\alpha}^{-1}(\text{spec } \mathbf{K}_\alpha)\,]' \cup D'\,)'$$

$$\subset [\,\hat{\alpha}^{-1}(\text{spec } \mathbf{K}_\alpha)\,]'' \cup [\,\hat{\alpha}^{-1}(\{0\})\,]' = \phi \cup \phi = \phi\,. \quad \square$$

We state one more corollary of Proposition 4.4.1 for later reference.

Corollary 3. If α_1 and α_2 are two smooth compactly supported functions on G such that $\alpha_1(k^{-1}gk) = \alpha_1(g)$ and $\alpha_2(k^{-1}gk) = \alpha_2(g)$ for all $g \in G$, $k \in K$, and $\hat{\alpha}_1(\Lambda)$, $\hat{\alpha}_2(\Lambda)$ are nonconstant, then

$$[\,a_C^* \setminus \hat{\alpha}_1^{-1}(\text{spec } \mathbf{K}_{\alpha_1}) \cup D_1\,] \cap [\,a_C^* \setminus \hat{\alpha}_2^{-1}(\text{spec } \mathbf{K}_{\alpha_2}) \cup D_2\,]$$

is connected.

Proof. We have

$$[\,a_C^* \setminus \hat{\alpha}_1^{-1}(\text{spec } \mathbf{K}_{\alpha_1}) \cup D_1\,] \cap [\,a_C^* \setminus \hat{\alpha}_2^{-1}(\text{spec } \mathbf{K}_{\alpha_2}) \cup D_2\,]$$

$$= a_C^* \setminus [\,\hat{\alpha}_1^{-1}(\text{spec } \mathbf{K}_{\alpha_1}) \cup D_1\,] \cup [\,\hat{\alpha}_2^{-1}(\text{spec } \mathbf{K}_{\alpha_2}) \cup D_2\,]\,,$$

where

$$\{ [\,\hat{\alpha}_1^{-1}(\text{spec }\mathbf{K}_{\alpha_1}) \cup D_1\,] \cup [\,\hat{\alpha}_2^{-1}(\text{spec }\mathbf{K}_{\alpha_2}) \cup D_2\,] \}''$$

$$= [\,\hat{\alpha}_1^{-1}(\text{spec }\mathbf{K}_{\alpha_1}) \cup D_1\,]'' \cup [\,\hat{\alpha}_2^{-1}(\text{spec }\mathbf{K}_{\alpha_2}) \cup D_2\,]''$$

$$= \phi \cup \phi = \phi\,. \;\Box$$

As claimed, $\hat{\alpha}^{-1}(\text{spec }\mathbf{K}_\alpha) \cup D$ is a countable closed set. Since

$$(\boldsymbol{a}_{\mathrm{C}}^*)^+ \cap [\,\boldsymbol{a}_{\mathrm{C}}^* \setminus \hat{\alpha}^{-1}(\text{spec }\mathbf{K}_\alpha) \cup D\,] = (\boldsymbol{a}_{\mathrm{C}}^*)^+ \setminus \hat{\alpha}^{-1}(\text{spec }\mathbf{K}_\alpha) \cup D\,,$$

which is connected by Corollary 2, and $\phi(\mathrm{J}')$ and $\mathrm{D}_{\mathrm{J}'}/\mathrm{D}$ agree on some open subset of $(\boldsymbol{a}_{\mathrm{C}}^*)^+ \setminus \hat{\alpha}^{-1}(\text{spec}\mathbf{K}_\alpha) \cup D$, each $\phi(\mathrm{J}')$, originally defined on $(\boldsymbol{a}_{\mathrm{C}}^*)^+$, can be analytically continued to a holomorphic function on

$$(\boldsymbol{a}_{\mathrm{C}}^*)^+ \cup [\,\boldsymbol{a}_{\mathrm{C}}^* \setminus \hat{\alpha}^{-1}(\text{spec }\mathbf{K}_\alpha) \cup D\,] = \boldsymbol{a}_{\mathrm{C}}^* \setminus \hat{\alpha}^{-1}(\text{spec }\mathbf{K}_\alpha) \cup D\,.$$

Let us examine the nature of its singularities. Since later on we will introduce another kernel function, we write $\alpha_1, \mathbf{K}_1, D_1$ in place of $\alpha, \mathbf{K}_\alpha, D$, and denote the extension of $\phi(\mathrm{J}')$ by $\phi^{(1)}(\mathrm{J}')$.

We remind ourselves that $D_1 \subset \boldsymbol{a}_{\mathrm{C}}^* \setminus \hat{\alpha}_1(\text{spec }\mathbf{K}_1)$. Since each point of D_1 is isolated, the points of D_1 are zeros of D of finite order, so $\phi^{(1)}(\mathrm{J}')$ has at worst a pole at a point of D_1.

Suppose $\Lambda_0 \in \hat{\alpha}_1^{-1}(\text{spec }\mathbf{K}_1)$. Consider first the easy case when $\Lambda_0 \in \hat{\alpha}_1^{-1}(\text{spec }\mathbf{K}_1 \setminus \{0\})$. The function $\mathrm{D}_{\mathrm{J}'}$ has at worse a pole at Λ. So does D. In order that $\phi^{(1)}(\mathrm{J}')$ has no worse than a pole at Λ_0, it needs only that Λ_0, in case $\mathrm{D}(\Lambda_0) = 0$, is a zero of D of finite order. Since $\Lambda_0 \notin \hat{\alpha}_1^{-1}(\{0\})$, there is a deleted neighborhood of Λ_0 which lies entirely in $\boldsymbol{a}_{\mathrm{C}}^* \setminus \hat{\alpha}_1^{-1}(\text{spec }\mathbf{K}_\alpha)$. Then, in view of Corollary 1, it is clear that D cannot have a zero of infinite order at Λ_0.

We now consider the case when $\Lambda_0 \in \hat{\alpha}_1^{-1}(\{0\})$. Our aim is to show that $\phi^{(1)}(\mathrm{J}')$, apart from removable singularities, has at worse a pole at Λ_0. From Theorem 2.2.1, we know that there exists another smooth compactly supported function α_2 on G such that $\alpha_2(k^{-1}gk) = \alpha_2(g)$ for all $g \in G$, $k \in K$, and $\hat{\alpha}_2(\Lambda_0) \neq 0$. If α_2 replaces α_1 in all of the preceding development, we get that the original $\phi(J')$ can be analytically continued to a holomorphic function $\phi^{(2)}(J')$ on $\boldsymbol{a}_C^* \setminus \hat{\alpha}_2^{-1}(\text{spec } \mathbf{K}_2) \cup D_2$. Moreover, because $\Lambda_0 \notin \hat{\alpha}_2^{-1}(\{0\})$, $\phi^{(2)}(J')$ has at worse a pole at Λ_0. Let H denote

$$[\, \boldsymbol{a}_C^* \setminus \hat{\alpha}_1^{-1}(\text{spec } \mathbf{K}_1) \cup D_1 \,] \cap [\, \boldsymbol{a}_C^* \setminus \hat{\alpha}_2^{-1}(\text{spec } \mathbf{K}_2) \cup D_2 \,]$$

$$= \boldsymbol{a}_C^* \setminus [\, \hat{\alpha}_1^{-1}(\text{spec } \mathbf{K}_1) \cup D_1 \,] \cup [\, \hat{\alpha}_2^{-1}(\text{spec } \mathbf{K}_2) \cup D_2 \,] .$$

Of course,

$$[\, \hat{\alpha}_1^{-1}(\text{spec } \mathbf{K}_1) \cup D_1 \,] \cup [\, \hat{\alpha}_2^{-1}(\text{spec } \mathbf{K}_2) \cup D_2 \,]$$

is a countable closed set. Thus $H \cap (\boldsymbol{a}_C^*)^+$ is a nonempty open subset of H on which $\phi^{(1)}(J')$ and $\phi^{(2)}(J')$ agree. By Corollary 3, H is connected, so $\phi^{(1)}(J')$ and $\phi^{(2)}(J')$ agree on all of H. Now because $\Lambda_0 \notin \hat{\alpha}_2^{-1}(\{0\})$, there exists a deleted neighborhood N of Λ_0 such that

$$\mathrm{N} \cap [\, \hat{\alpha}_2^{-1}(\text{spec } \mathbf{K}_2) \cup D_2 \,] = \phi .$$

The function $\phi^{(2)}(J')$ is holomorphic on N. Since the set of limit points of $\hat{\alpha}_1^{-1}(\text{spec } \mathbf{K}_1) \cup D_1$ is discrete, we may assume that N has no limit points of $\hat{\alpha}_1^{-1}(\text{spec } \mathbf{K}_1) \cup D_1$. The set

$$\mathrm{N} \setminus [\, \hat{\alpha}_1^{-1}(\text{spec } \mathbf{K}_1) \cup D_1 \,] \cup [\, \hat{\alpha}_2^{-1}(\text{spec } \mathbf{K}_2) \cup D_2 \,] ,$$

on which $\phi^{(1)}(J')$ and $\phi^{(2)}(J')$ agree, equals

$$\mathrm{N} \setminus \mathrm{N} \cap [\, \hat{\alpha}_1^{-1}(\text{spec } \mathbf{K}_1) \cup D_1 \,] .$$

But each point x of $\mathrm{N} \cap [\, \hat{\alpha}_1^{-1}(\text{spec } \mathbf{K}_1) \cup D_1 \,]$ is isolated. Therefore, the holomorphicity of $\phi^{(2)}(J')$ at x implies that x is a removable singularity of $\phi^{(1)}(J')$. We can then have $\phi^{(1)}(J' \mid \Lambda) = \phi^{(2)}(J' \mid \Lambda)$ for all $\Lambda \in \mathrm{N}$, and it follows immediately that $\phi^{(1)}(J')$ has at worst a pole at Λ_0. We have just arrived at one of the main results of this work:

Theorem 4.4.1. *The holomorphic functions* $\phi(J)$, *originally defined on* $(\boldsymbol{a}_C^*)^+$, *can be analytically continued to meromorphic functions on the entire* $\boldsymbol{a}_C^*$.

We shall always denote these extensions, which are unique by the identity theorem of holomorphic functions, by $\phi(J)$ also.

We next turn to consider the analytic continuation of the Eisenstein series itself.

Theorem 4.4.2. *For each fixed* $g \in G$, *the Eisenstein series, originally defined on* $(\boldsymbol{a}_C^*)^+$, *can be analytically continued to a meromorphic function on the entire* $\boldsymbol{a}_C^*$. *Moreover,* S, *the union of the sets of poles of* $E(\Lambda, g)$ *as* g *varies over* G, *has no limit point.*

Proof. Let us begin with a choice of the kernel function, α_1. Suppose at first that g is a fixed element of G. Recall that $F_J^{*(1)}(\Lambda, g)$ are holomorphic functions of $\boldsymbol{a}_C^* \setminus \hat{\alpha}_1^{-1}(\text{spec } \mathbf{K}_1)$ which have, depending on g perhaps, either a removable singularity or a pole at each point of $\hat{\alpha}_1^{-1}(\text{spec } \mathbf{K}_1 \setminus \{0\})$. Referring to the development for Theorem 4.4.1, we know that $\sum_j \phi(J)F_J^{*(1)}$ defines a holomorphic function on $\boldsymbol{a}_C^* \setminus \hat{\alpha}_1^{-1}(\text{spec } \mathbf{K}_1) \cup D_1$. By Corollary 2 of Proposition 4.4.1, the set $(\boldsymbol{a}_C^*)^+ \cap [\, \boldsymbol{a}_C^* \setminus \hat{\alpha}_1^{-1}(\text{spec } \mathbf{K}_1) \cup D_1 \,]$, which is equal to $(\boldsymbol{a}_C^*)^+ \setminus \hat{\alpha}_1^{-1}(\text{spec } \mathbf{K}_1) \cup D_1$, is connected. Now recall the expression (3.5.4)

$$E(\Lambda) = \sum_J \phi(J \mid \Lambda) F_J^{*(1)}(\Lambda) \qquad \text{for all } \Lambda \in (\boldsymbol{a}_C^*)^+ \setminus \hat{\alpha}_1^{-1}(\text{spec } \mathbf{K}_1)\,.$$

We can extend $E(\Lambda)$ to a holomorphic function $E^{(1)}(\Lambda)$ on $(\boldsymbol{a}_C^*)^+ \cup [\, \boldsymbol{a}_C^* \setminus \hat{\alpha}_1^{-1}(\text{spec } \mathbf{K}_1) \cup D_1 \,]$, which contains $\boldsymbol{a}_C^* \setminus \hat{\alpha}_1^{-1}(\text{spec } \mathbf{K}_1) \cup D_1$, by simply setting:

$$E(\Lambda, g) = \sum_J \phi(J \mid \Lambda) F_J^{*(1)}(\Lambda, g) \qquad \text{for all } \Lambda \in \boldsymbol{a}_C^* \setminus \hat{\alpha}_1^{-1}(\text{spec } \mathbf{K}_1) \cup D_1 \qquad (g \in G)\,.$$

If $\Lambda_0 \in [\, \hat{\alpha}_1^{-1}(\text{spec } \mathbf{K}_1) \cup D_1 \,] \setminus \hat{\alpha}_1^{-1}(\{0\})$, there is a deleted neighborhood of Λ_0 which is con-

$E^{(1)}(\Lambda)$ has, depending on g perhaps, either a removable singularity or a pole.

To make way for the second assertion of the theorem, we remark that for any $\Lambda_0 \in \mathfrak{a}_C^* \setminus \alpha_1^{-1}(\{0\})$ there is a deleted neighborhood U of Λ_0 on which, by definition,

$$E^{(1)}(\Lambda, g) = \sum_J \phi(J \mid \Lambda) F_J^{*(1)}(\Lambda, g) \qquad \text{for all } g \in G.$$

Referring to Corollary 1 to Theorem 3.6.1, we may even assume that for all $g \in G$, the functions $F_J^{*(1)}(\Lambda, g)$ are holomorphic on U. We may also assume that $\phi(J \mid \Lambda)$ too are holomorphic on U. For all $g \in G$, $E(\Lambda, g)$ is thus holomorphic on U.

Returning to the proof of the first assertion, suppose on the other hand that $\Lambda_0 \in \hat{\alpha}_1^{-1}(\{0\})$. Introduce another kernel function α_2 such that $\hat{\alpha}_2(\Lambda_0) \neq 0$, as in the development for Theorem 4.4.1. The preceding then shows that

$$E^{(2)}(\Lambda, g) = \sum_J \phi(J \mid \Lambda) F_J^{*(2)}(\Lambda, g)$$

has either a regular point or a pole at Λ_0. By an argument employed in the continuation of $\phi(J')$, for the fixed g, $E^{(1)}(\Lambda, g) = E^{(2)}(\Lambda, g)$ in some deleted neighborhood of Λ_0, so $E^{(1)}$ has either a removable singularity or a pole at Λ_0. This completes the proof of the first assertion.

For each $g \in G$, the meromorphic extension to the entire $\mathfrak{a}_C^*$ will again be denoted by $E(\Lambda, g)$.

If instead of α_1 we use α_2 to begin the analytic construction of the Eisenstein series, we will, by virtue of the identity theorem of holomorphic functions, arrive at the same function. Because of this it can be claimed as well that for any $\Lambda_0 \in \mathfrak{a}_C^* \setminus \hat{\alpha}_2^{-1}(\{0\})$ there is a deleted neighborhood U of Λ_0 on which $E(\Lambda, g) = \sum_J \phi(J \mid \Lambda) F_J^{*(2)}(\Lambda, g)$, this implies also that for all $g \in G$, $E(\Lambda, g)$ is holomorphic on U. Therefore, for any $\Lambda_0 \in \mathfrak{a}_C^*$, there is a deleted neighborhood U of Λ_0 on which $E(\Lambda, g) = \sum_J \phi(J \mid \Lambda) F_J^*(\Lambda, g)$ $(g \in G)$ for some choice of kernel function α; in particular, $E(\Lambda, g)$ is holomorphic on U for each $g \in G$ and $E(\Lambda, g) \in C^\infty(U \times G)$. It is immediate that S has no limit points. For if Λ_0 were a limit point of S, then there could not be a deleted neighborhood of Λ_0 on which $E(\Lambda, g)$ is holomorphic for all $g \in G$. This completes the proof of Theorem 4.4.2. □

We continue to denote the set of all poles of $E(\Lambda, g)$ by S.

The smoothness of $E(\Lambda, g)$ on $\mathfrak{a}_C^* \setminus S \times G$ is our next concern. Before that we prove a little technical lemma.

Lemma 4.4.1. Let $z_0 \in \mathbf{C}$ and N be a deleted neighborhood of z_0, $f(z, g) \in C^\infty(N \times G)$ and be holomorphic on N for each $g \in G$. Suppose $f(z, g)$ has a removable singularity at z_0 for every $g \in G$, then it belongs to $C^\infty(N \cup \{z_0\} \times G)$.

Proof. Let β be a positively oriented circle in N whose interior forms a neighborhood at z_0. Fix another, smaller, disc neighborhood U of z_0. Since for each $g \in G$, $f(z, g)$ is actually holomorphic on $N \cup \{z_0\}$, Cauchy's formula gives

$$f(z, g) = \frac{1}{2\pi i} \int_\beta \frac{f(u, g)}{u-z}\, du \qquad ((z, g) \in U \times G).$$

As $f(u, g)/u-z$ is continuous on $\beta^* \times U \times G$, it is easily verified that $f(z, g) \in C(U \times G)$. Also, if $V \subset \mathbf{R}^n \to W \subset G : (t_1, \ldots, t_n) \mapsto g(t_1, \ldots, t_n)$ is a coordinate mapping,

$$\frac{\partial^{\lambda_1 + \cdots + \lambda_n} f}{\partial^{\lambda_1} t_1 \cdots \partial^{\lambda_n} t_n}(z, g(t, \ldots, t_n))$$

$$= \frac{1}{2\pi i} \int_\beta \frac{1}{u-z} \frac{\partial^{\lambda_1 + \cdots + \lambda_n} f}{\partial^{\lambda_1} t_1 \cdots \partial^{\lambda_n} t_n}(u, g(t_1, \ldots, t_n))\, du$$

$((z, g(t_1, \ldots, t_n)) \in U \times W)$. Therefore, likewise

$$\frac{\partial^{\lambda_1+\cdots+\lambda_n} f}{\partial^{\lambda_1} t_1 \cdots \lambda^{\lambda_n} t_n}(z, g) \in C(U \times W), \qquad (\lambda_1, \ldots, \lambda_n) \in (\mathbf{Z}^+)^n .$$

To complete the proof, it suffices to show that

$$\frac{d^\nu f}{dz^\nu}(z, g) \in C(U \times G), \quad \nu \in \mathbf{Z}^+ .$$

For this we need only recall that

$$\frac{d^\nu f}{dz^\nu}(z, g) = \frac{(\nu-1)!}{2\pi i} \int_\beta \frac{f(u, g)}{(u-z)^\nu}\, du \qquad ((z, g) \in U \times G). \qquad \square$$

Theorem 4.4.3. The meromorphic extension of the Eisenstein series, $E(\Lambda, g)$, is in $C^\infty(\mathfrak{a}_C^* \setminus S \times G)$.

Proof. Once again we parametrize $\mathfrak{a}_C^*$ by a complex variable and thus identify $\mathfrak{a}_C^*$ with $\mathbf{C}$. Let $z_0 \in \mathfrak{a}_C^* \setminus S$. There is a deleted neighborhood U of z_0 on which $E(z, g) = \sum_J \phi(J \mid z) F_J^*(z, g)$ $(g \in G)$ for some choice of kernel function α, and $\hat{\alpha}(z) \neq 0$ for $z \in \tilde{U}$, where $\tilde{U} = U \cup \{z_0\}$.

On the other hand, since $z_0 \notin S$, $E(z, g)$ is holomorphic on $\tilde{U}$ for each $g \in G$. Also, it belongs to $C^\infty(U \times G)$. We have

$$E(z, g) = \sum_{k=0}^{\infty} (z - z_0)^k a_k(g) \qquad ((z, g) \in \tilde{U} \times G),$$

where

$$a_k(g) = (2\pi i)^{-1} \int_\beta (u - z_0)^{-k-1} E(u, g)\, du ,$$

β being a positively oriented circle in U of radius r and centered at z_0. It is clear that $a_k \in C(G)$. If Ω is any compact subset of G, there exists a constant M such that

$$|a_k(g)| \leq \frac{M}{r^k} \qquad (g \in \Omega) \quad \text{for all } k .$$

Hence there is a neighborhood N of z_0 such that the infinite series converges uniformly on $N \times \Omega$. This implies that $E(z, g) \in C(N \times \Omega)$. The same can be said of any derivative of E(z, g) with respect to z, $(d^{\nu}E/dz^{\nu})(z, g)$. Since Ω is arbitrary, we conclude that E(z, g) and all its derivatives with respect to z belong to $C(N \times G)$, and hence to $C(\tilde{U} \times G)$, as they are already in $C^{\infty}(U \times G)$.

Now $E(z, g) = \sum_J \phi(J \mid z)F_J^*(z, g)$ implies that E(z, g) satisfies the Fredholm equation:

$$(\mathbf{K}_\alpha - \hat{\alpha}(z))E(z, g) = -\sum_J \hat{\alpha}(z)\phi(J \mid z)H_J(z, g) \qquad ((z, g) \in U \times G).$$

By Proposition 3.3.3, the left-hand side of the equation is defined and continuous on $\tilde{U} \times G$. Therefore, for every $g \in G$, the right-hand side of the equation has a removable singularity at z_0. Lemma 4.4.1 implies that $-\sum_J \hat{\alpha}(z)\phi(J \mid z)H_J(z, g)$ is actually in $C^{\infty}(\tilde{U} \times G)$. Then, together with Proposition 3.3.3,

$$E(z, g) = \frac{K_\alpha E(z, g) + \sum_J \hat{\alpha}(z)\phi(J \mid z)H_J(z, g)}{\hat{\alpha}(z)} \qquad ((z, g) \in \tilde{U} \times G)$$

implies that $E(z, g) \in C^{\infty}(\tilde{U} \times G)$. Since $z_0 \in \tilde{U}$, and z_0 is an arbitrary point of $\boldsymbol{a}_{\mathrm{C}}^* \setminus S$, the theorem has been proved. □

The continued Eisenstein series has the same important growth property as the original.

Theorem 4.4.4. For each $\Lambda_0 \in \boldsymbol{a}_{\mathrm{C}}^* \setminus S$, $E(\Lambda_0, g)$ is a slowly increasing function on G.

Proof. We identify $\boldsymbol{a}_{\mathrm{C}}^*$ with **C**, and use the notation in the proof of the last theorem. Let $z_0 \in \boldsymbol{a}_{\mathrm{C}}^* \setminus S$. We have

$$E(z_0, g) = (2\pi i)^{-1} \int_\beta (u - z_0)^{-1} E(u, g)\, du$$

$$= (2\pi i)^{-1} \int_{\beta} (u - z_0)^{-1} \sum_{J} \phi(J\,|\,u) F_J^{*}(u, g)\, du$$

$$= (2\pi i)^{-1} \int_{\beta} (u - z_0)^{-1} \sum_{J} \phi(J\,|\,u)(F_J^{**}(u, g) + H_J(u, g))\, du$$

$$= \sum_{J} (2\pi i)^{-1} \int_{\beta} (u - z_0)^{-1} \phi(J\,|\,u) F_J^{**}(u, g)\, du$$

$$+ \sum_{J} (2\pi i)^{-1} \int_{\beta} (u - z_0)^{-1} \phi(J\,|\,u) H_J(u, g)\, du\,.$$

By Corollary 2 to Theorem 3.6.1, the first sum is bounded over G. As to the second sum, let us recall from Section 3.4 that there exist $C_1, C_2 > 0$ such that

$$|H_J^{o}(u, g)| \le C_1 e^{-C_2\eta(H_0(g))} \qquad \text{for } (u, g) \in \beta^* \times S_0\,.$$

Because $H_J(u, g)$ is simply $H_J^{o}(u, g)/(\hat{\alpha}^{o}(u))^2$, the second sum is obviously slowly increasing. This completes the proof of Theorem 4.4.4. □

The continued Eisenstein series inherited other characteristic properties of the original as well. We list them in a series of propositions. There we mostly regard $E(\Lambda, g)$ as defined on $\mathfrak{a}_C^* \setminus S \times G$.

Proposition 4.4.2. For every $\Lambda \in \mathfrak{a}_C^* \setminus S$, the continued Eisenstein series is a Γ-automorphic function on G, i.e., $E(\Lambda, \gamma g) = E(\Lambda, g) \quad (g \in G)$ for all $\gamma \in \Gamma$.

Proof. For each $\gamma \in \Gamma$ and $g \in G$, both sides of the equation are meromorphic functions on the entire $\mathfrak{a}_C^*$, holomorphic on $\mathfrak{a}_C^* \setminus S$; and since they agree on a nonempty open set, namely $(\mathfrak{a}_C^*)^+$, the original domain of definition of the Eisenstein series, they agree on $\mathfrak{a}_C^* \setminus S$. □

Proposition 4.4.3. For every $\Lambda \in \mathfrak{a}_C^* \setminus S$, the continued Eisenstein series is a σ-function on G, i.e., $E(\Lambda, gk) = \sigma(k) E(\Lambda, g) \quad (g \in G)$ for all $k \in K$.

Proof. The proof is completely analogous to that of Proposition 4.4.2. □

Proposition 4.4.4. For every $\Lambda \in \mathfrak{a}_C^* \setminus S$, the continued Eisenstein series is an eigenfunction for $\mathbf{Z}(G)$, thus $DE(\Lambda) = \chi_\Lambda(D)E(\Lambda)$ for all $D \in \mathbf{Z}(G)$.

Proof. It suffices to show that for each $g \in G$, $DE(\Lambda, g)$ is holomorphic on $\mathfrak{a}_C^* \setminus S$. Identify $\mathfrak{a}_C^*$ with $\mathbf{C}$. Let $z_0 \in \mathfrak{a}_C^* \setminus S$ and U be a disc neighborhood of z_0 contained in $\mathfrak{a}_C^* \setminus S$. We shall apply Morera's theorem of complex analysis. Let Δ be a triangle in U, we need to show that

$$\int_{\partial\Delta} DE(z, g)\, dz = 0 \qquad (g \in G)\,.$$

Of course,

$$\int_{\partial\Delta} E(Z, g)\, dz = 0\,.$$

So it is a matter of verifying that

$$D \int_{\partial\Delta} E(z, g)\, dz = \int_{\partial\Delta} DE(z, g)\, dz\,.$$

This poses no difficulty because $\partial\Delta$ is compact and $E(z, g) \in C^\infty(U \times G)$. □

Proposition 4.4.5. If P_i is a maximal standard cuspidal subgroup associate to P, $P_i = N_i A_i M_i$, then

$$\int_{N_i \cap \Gamma \setminus N_i} E(\Lambda, ng)\, dn = \sum_{s,n} \phi(i, s, n \mid \Lambda) e^{(-s\Lambda - \rho_i)(\log a_i)} \sigma(k) \Psi_{i,s,n}(m_i) \quad (g = n_i a_i m_i k)\,.$$

Proof. For each $g \in G$, although the right-hand side of the equation may not be holomorphic on $\mathfrak{a}_C^* \setminus S$, it is nevertheless a meromorphic function on the entire $\mathfrak{a}_C^*$. To show that

$$\int_{N_i \cap \Gamma \setminus N_i} E(\Lambda, ng)\, dn$$

is meromorphic in the entire $\mathfrak{a}_C^*$ one can apply Morera's theorem together with Fubini's. □

The proof of the next proposition is not much different.

Proposition 4.4.6. If P_l is a maximal standard cuspidal subgroup not associate to P, $P_l = N_l A_l M_l$, then

$$\int_{N_l \cap \Gamma \backslash N_l} E(\Lambda, ng)\, dn = 0 .$$

CHAPTER 5

FUNCTIONAL EQUATIONS

5.1 Introduction

For each $\Phi \in {}^0L^2(\Gamma_M \backslash M, \sigma_M, \chi)$ the entire meromorphicity of the constant term coefficients $\phi(j, s, n \mid \Lambda)$ and the Eisenstein series $E(\Lambda, \Phi, g)$ itself have been established. It is now a simple matter to derive the functional equation:

$$E(\Lambda, \Phi, g) = E(s\Lambda, M(s \mid \Lambda)\Phi, g) \qquad (\Lambda \in \boldsymbol{a}_C^*) ;$$

equivalently,

$$E(s^{-1}\Lambda_i , \Phi, g) = E(\Lambda_i , M(s \mid s^{-1}\Lambda_i)\Phi, g) \qquad (\Lambda_i \in (\boldsymbol{a}_i)_C^*) . \tag{5.1.1}$$

Both sides of this last equation possess the characteristics of an Eisenstein series, and therefore the uniqueness development of Section 4.3 is applicable here. Of course, here as in Section 4.3, the combinatorial structure of their constant terms is very simple. The functional equation of the intertwining operator $M(s, \Lambda)$ is:

$$M(s^{-1} \mid s\Lambda)M(s \mid \Lambda) = I \qquad (\Lambda \in \boldsymbol{a}_C^*) ;$$

equivalently,

$$M(s^{-1} \mid \Lambda_1)M(s \mid s^{-1}\Lambda_i) = I \qquad (\Lambda_i \in (\boldsymbol{a}_i)_C^*) ,$$

which will be obtained by comparing the constant terms of the two sides of the equation (5.1.1).

5.2. Functional Equations

Let P_i be a maximal standard cuspidal subgroup associated with P. Let s be an element of $W(\boldsymbol{a}, \boldsymbol{a}_i)$.

With the meromorphicity of the constant term coefficients $\phi(j, s, n \mid \Lambda)$ and the Eisenstein series $E(\Lambda, g)$ itself fully established, we proceed to derive the functional equations of $E(\Lambda, g)$. Referring to Chapter 4, for a given choice of kernel function α, the poles of $\phi(j, s, n \mid \Lambda)$ and $E(\Lambda, g)$ all lie in $\hat{\alpha}^{-1}(\text{spec}\,\mathbf{K}_\alpha) \cup D$. It is clear that for a fixed maximal standard cuspidal subgroup P, $\hat{\alpha}^{-1}(\text{spec}\,\mathbf{K}_\alpha) \cup D$ depends only on χ, if the cusp form Φ is used to construct the Eisenstein series is in ${}^oL^2(\Gamma_M \backslash M, \sigma_M, \chi)$. Recall from Section 3.2 the relationship between $\phi(j, s, n \mid \Lambda)$ and $M(s \mid \Lambda)$, the intertwining operator between ${}^oL^2(\Gamma_M \backslash M, \sigma_M, \chi)$ and ${}^oL^2(\Gamma_{M_i} \backslash M, \sigma_{M_i}, {}^s\chi)$ that appears in Langlands' formula for the constant term of the Eisenstein series $E(\Lambda, \Phi, g)$ with respect to P_i : if $\{\Phi_p\}_p$ and $\{\Psi_{i,s,n}\}_n$ are respectively bases of ${}^oL^2(\Gamma_M \backslash M, \sigma_M, \chi)$ and ${}^oL^2(\Gamma_{M_i} \backslash M, \sigma_{M_i}, {}^s\chi)$, and $M(s \mid \Lambda)$ has the matrix representation $[\theta(i, s, n; p \mid \Lambda)]$, then for

$$\Phi = \sum_p \eta_p \Phi_p ,$$

we have

$$M(s \mid \Lambda)\Phi = \sum_n \phi(i, s, n \mid \Lambda)\Psi_{i,s,n} ,$$

where

$$\phi(i, s, n \mid \Lambda) = \sum_p \eta_p \theta(i, s, n; p \mid \Lambda) .$$

Since the functions Φ_p share the same set of eigenvalues for $\mathbf{Z}(M)$, it can be established as before that each of $\theta(i, s, n; p \mid \Lambda)$ can be analytically continued to a meromorphic function on the entire $\mathfrak{a}_C^*$, whose set of poles is contained in $\hat{\alpha}^{-1}(\text{spec}\,\mathbf{K}_\alpha) \cup D$, which is common to all $\theta(i, s, n; p \mid \Lambda)$. In other words, each intertwining operator $M(s \mid \Lambda)$, originally defined and holomorphic on $(\mathfrak{a}_C^*)^+$, can be analytically continued to become meromorphic on $\mathfrak{a}_C^*$, whose set of poles is contained in $\hat{\alpha}^{-1}(\text{spec}\,\mathbf{K}_\alpha) \cup D$.

We shall prove that, for $\Phi \in {}^{0}L^2(\Gamma_M \backslash M, \sigma_M, \chi)$, $s \in W(\boldsymbol{a}, \boldsymbol{a}_i)$, and any fixed $g \in G$,

$$E(\Lambda, \Phi, g) = E(s\Lambda, M(s \mid \Lambda)\Phi, g) \qquad (\Lambda \in \boldsymbol{a}_C^*)\,; \tag{5.2.1}$$

equivalently,

$$\sum_p \eta_p E(\Lambda, \Phi_p, g) = \sum_n \phi(i, s, n \mid \Lambda) E(s\Lambda, \Psi_{i,s,n}, g) \qquad (\Lambda \in \boldsymbol{a}_C^*)\,. \tag{5.2.1$'$}$$

We shall also deduce that

$$M(s^{-1} \mid s\Lambda) M(s \mid \Lambda) = I \qquad (\Lambda \in \boldsymbol{a}_C^*)\,. \tag{5.2.2}$$

For convenience use $\tilde{S}$ to stand for $\hat{\alpha}^{-1}(\text{spec}\, \mathbf{K}_\alpha) \cup D$. The Eisenstein series $E(\Lambda_i, \Psi_{i,s,n}, g)$ can already be regarded as meromorphic functions on the entire $(\boldsymbol{a}_i)_C^*$, whose sets of poles are contained in a common countable closed set with a discrete set of limit points, $\tilde{S}_i$. The right-hand side of the equation (5.2.1) is therefore a meromorphic function on the entire $(\boldsymbol{a}_i)_C^*$ whose set of poles is contained in $\tilde{S} \cup s^{-1}(\tilde{S}_i)$. Implicit in the equation (5.2.1) or (5.2.1′) is the statement that $s^{-1}(\tilde{S}_i) \setminus \tilde{S}$ actually consists of removable singularities.

We now begin the proof of the functional equations. As discussed in Section 4.3., in the case $i = 1$, $\boldsymbol{a}_i = \boldsymbol{a}$ and $s = 1 \in W(\boldsymbol{a}, \boldsymbol{a})$, $M(1, \Lambda) =$ identity. Here the corresponding functional equations are trivially true. The only nontrivial case is when $s \in W(\boldsymbol{a}, \boldsymbol{a}_i)$ with $sH = -H_i$ ($i = 1$ or 2). We shall prove: for any fixed $g \in G$,

$$E(s^{-1}\Lambda_i, \Phi, g) = E(\Lambda_i, M(s \mid s\Lambda_i)\Phi, g) \qquad (\Lambda_i \in (\boldsymbol{a}_i)_C^*) \tag{5.2.3}$$

which is obviously equivalent to the equation (5.2.1); and

$$M(s^{-1} \mid \Lambda_i) M(s \mid s^{-1}\Lambda_i) = I \qquad (\Lambda \in (\boldsymbol{a}_i)_C^*) \tag{5.2.4}$$

which is obviously equivalent to the equation (5.2.2). For convenience, introduce $E_1(\Lambda_i, g)$ and $E_2(\Lambda_i, g)$ to denote respectively the left-hand side and right-hand side of the equation (5.2.3); these are meromorhpic functions on the entire $(\boldsymbol{a}_i)_C^*$. To establish the equality of two meromorphic functions, it suffices to prove it over a nonempty open set that avoids all the singularities. the equality then holds wherever one of the two sides is regular. In terms of the basis $\{\Psi_{i,s,n}\}_n$,

$$E_2(\Lambda_i, g) = E(\Lambda_i, M(s\, s^{-1}\Lambda_i)\Phi, g) = \sum \phi(i, s, n \mid s^{-1}\Lambda_i) E(\Lambda_i, \Psi_{i,s,n}, g) .$$

For a fixed $\Lambda_i \in ((\boldsymbol{a}_i)^*_{\mathbb{C}})^+$, each $E(\Lambda_i, \Psi_{i,s,n}, g)$, being an Eisenstein series, is Γ-automorphic, slowly increasing, a σ-function and an eigenfunction for all of $\mathbf{Z}(G)$. Let $\Lambda_i \in ((\boldsymbol{a}_i)^*_{\mathbb{C}})^+$ be such that $s^{-1}\Lambda_i$ is not a pole of $E(\Lambda, \Phi, g)$ for all $g \in G$ or of any of $\phi(i, s, n \mid \Lambda)$. Then $E_2(g, \Lambda_i)$ is Γ-automorphic, slowly increasing, a σ-function and an eigenfunction for all of $\mathbf{Z}(G)$. It is part of the conclusion of the last chapter than $E_1(g, \Lambda_i)$ has the same properties as well. Then, by Theorem 3.5.2,

$$E_1(g) - \sum_r [\, \delta_r E_1^r \,]_F (g)$$

and

$$E_2(g) - \sum_r [\, \delta_r E_2^r \,]_F (g)$$

are both bounded on G.

Now

$$E_2(g) - E_1(g) = (\, E_2(g) - \sum_r [\, \delta_r E_2^r \,]_F (g)\,) - (\, E_1(g) - \sum_r [\, \delta_r E_1^r \,]_F (g)\,)$$
$$+ \sum_r [\, \delta_r E_2^r \,]_F (g) - \sum_r [\, \delta_r E_1^r \,]_F (g)$$

Let us consider the term

$$\sum_r [\, \delta_r E_2^r \,]_F (g) - \sum_r [\, \delta_r E_1^r \,]_F (g) .$$

We have

$$E_2^r(g) = E^r(\Lambda_i, M(s \mid s^{-1}\Lambda_i), \Phi, g) = \sum_n \phi(i, s, n \mid s^{-1}\Lambda_i) E^r(\Lambda_i, \Psi_{i,s,n}, g).$$

Being an Eisenstein series for the maximal standard cuspidal subgroup P_i, which has as its associates only itself and P, $E^r(\Lambda_i, \Psi_{i,s,n}, g) = 0$ for $r \neq i$ or 1. (The index i is either 1 or 2. The first case is when P itself is its only associate, the second case is when P has another associate, P_2.) So $E_2^r(g) = 0$ for $r \neq i$ or 1, and

$$\sum_r [\, \delta_r E_2^r \,]_F (g) = [\, \delta_i e^{(-\Lambda_i - \rho_i)(\log a_i)} \sigma(k) M(s \mid s^{-1}\Lambda_i) \Phi(m_i) \,]_F$$

$$+ [\, \delta_1 e^{(-s^{-1}\Lambda_i - \rho)(\log a)} \sigma(k) M(s^{-1} \mid \Lambda_i) M(s \mid s^{-1}\Lambda_i) \Phi(m) \,]_F$$

($s^{-1} \in W(\boldsymbol{a}_i, \boldsymbol{a})$), where we have used Langlands' formula for constant terms for $E^r(\Lambda_i, M(s \mid s^{-1}\Lambda_i)\Phi, g)$. In view of the conclusion of the last chapter, it is also true that $E_1^r(g) = 0$ for $r \neq i$ or 1, whereas

$$\sum_r [\, \delta_r E_1^r \,]_F (g) = [\, \delta_1 e^{(-s^{-1}\Lambda_i - \rho)(\log a)} \sigma(k) \Phi(m) \,]_F$$

$$+ [\, \delta_i e^{(-s(s^{-1}\Lambda_i) - \rho_i)(\log a_i)} \sigma(k) M(s \mid s^{-1}\Lambda_i) \Phi(m_i) \,]_F$$

$$= [\, \delta_i e^{(-\Lambda_i - \rho_i)(\log a_i)} \sigma(k) M(s \mid s^{-1}\Lambda_i) \Phi(m_i) \,]_F$$

$$+ [\, \delta_1 e^{(-s^{-1}\Lambda_i - \rho)(\log a)} \sigma(k) \Phi(m) \,]_F .$$

Therefore,

$$\sum_r [\, \delta_r E_2^r \,]_F (g) - \sum_r [\, \delta_r E_1^r \,]_F (g) = [\, \delta_1 e^{(-s^{-1}\Lambda_i - \rho)(\log a)} \sigma(k) [\, M(s^{-1} \mid \Lambda_i) M(s \mid s^{-1}\Lambda_i) - I \,] \Phi(m) \,]_F .$$

Let $\log a = H(g) = xH$, $x \in \mathbf{R}$. Now

$$e^{(-s^{-1}\Lambda_i - \rho)(\log a)} = e^{(-s^{-1}\Lambda_i - \rho)(H(g))} = e^{(-s^{-1}\Lambda_i - \rho)(xH)}$$

$$= e^{x(-s^{-1}\Lambda_i)(H) - x\rho(H)}$$

$$= e^{x\Lambda_i(H_i) - x\rho(H)}$$

$$= e^{x\{[\Lambda_i(H_i)-\rho_i(H_i)]+[\rho_i(H_i)-\rho(H)]\}} .$$

Recall that $((\boldsymbol{a}_i)^*_C)^+$ includes all $\Lambda_i \in (\boldsymbol{a}_i)^*_C$ such that $(\mathrm{Re}\Lambda_i)(H_i) - \rho_i(H_i) > 0$. Thus, as in the proof of Lemma 4.3.6, for $\Lambda_i \in ((\boldsymbol{a}_i)^*_C)^+$ such that

$$[\mathrm{Re}(\Lambda_i)(H_i) - \rho_i(H_i)] + [\rho_i(H_i) - \rho(H)] > 0 ,$$

$E_2(g) - E_1(g)$ is bounded.

Then $E_2(g) - E_1(g)$ is a bounded Γ-automorphic eigenfunction of the Casimir operator which is K-finite and $\mathbf{Z}(G)$-finite. By means of Theorem 4.3.1, we will be able to deduce that $E_2(g) - E_1(g) = 0$, provided that Λ is appropriately restricted. First it must be verified that $E_1(g)$ and $E_2(g)$ share the same eigenvalue of the Casimir operator ω. Since $E_1(\Lambda_1, g)$ equals $E(s^{-1}\Lambda_i , \Phi, g)$, it has the same eigenvalue as $\Phi_{s^{-1}\Lambda_i}$, that is, as in Lemma 4.3.7, $\chi_{s^{-1}\Lambda_i}$. On the other hand, $E_2(\Lambda_i , g)$ equals $E(\Lambda_i , M(s \,|\, s^{-1}\Lambda_i)\Phi, g)$, where $M(s \,|\, s^{-1}\Lambda_i)\Phi \in {}^0L^2(\Gamma_{M_i} \backslash M_i , \sigma_{M_i} {}^s\chi)$, so it has the same eigenvalue as $(M(s\, s^{-1}\Lambda_i)\Phi)_{\Lambda_i}$, that is $({}^s\chi)_{\Lambda_i}$.

<u>Lemma 5.2.1.</u> With notation as above, we have

$$\chi_{s^{-1}\Lambda_i} = ({}^s\chi)_{\Lambda_i} .$$

(See Harish-Chandra [2, p. 88].)

Recall from Lemma 4.3.7 that

$$({}^s\chi)_{\Lambda_i}(\omega) = {}^s\chi(\omega_m) + \Lambda_i(H_i)^2 - \rho_i(H_i)^2 .$$

Now ${}^s\chi(\omega_m) + \Lambda_i(H_i)^2 - \rho_i(H_i)^2$ is clearly a nonconstant entire function of Λ_i. There certainly exists $\Lambda_i \in ((\boldsymbol{a}_i)^*_C)^+$ such that

$$[\mathrm{Re}(\Lambda_i)(H_i) - \rho_i(H_i)] + [\rho_i(H_i) - \rho(H)] > 0 ,$$

$s^{-1}\Lambda_i$ is not a pole of $E(\Lambda, \Phi, g)$ for all $g \in G$ or of any of $\phi_{i,s,n}(\Lambda)$ and ${}^s\chi(\omega_m) + \Lambda_i(H_i)^2 - \rho_i(H_i)^2$ is nonreal. Indeed there is a whole neighborhood N of Λ_i over which the conditions are satisfied. Then by Theorem 4.3.1, for $\Lambda_i \in N$,

$$E_1(\Lambda_i, g) = E_2(\Lambda_i, g) \qquad \text{for all } g \in G,$$

i.e.,

$$E_1(\Lambda_i, g) = E_2(\Lambda_i, g) \qquad \text{for all } (\Lambda_i, g) \in N \times G.$$

This implies the functional equation of the Eisenstein series.

Furthermore, if $E_1(\Lambda_i, g) = E_2(\Lambda_i, g)$ for all $g \in G$, where $\Lambda_i \in N$, a certain nonempty open set, then

$$0 = \sum_r E_2^r(g) - \sum_r E_1^r(g) = e^{(-s^{-1}\Lambda_i - \rho)(\log a)}\sigma(k)[\, M(s^{-1} \mid \Lambda_i)M(s \mid s^{-1}\Lambda_i) - I\,]\Phi(m)$$

for all $(\Lambda_i, g) \in N \times G$, which implies

$$[\, M(s^{-1} \mid \Lambda_i)M(s \mid s^{-1}\Lambda_i) - I\,]\Phi = 0 \qquad (\Lambda_i \in N)$$

as long as $\Phi \in {}^oL^2(\Gamma_M \backslash M, \sigma_M, \chi)$. Hence

$$M(s^{-1} \mid \Lambda_i)M(s \mid s^{-1}\Lambda_i) - I = 0 \qquad (\Lambda_i \in N).$$

It follows that

$$M(s^{-1} \mid \Lambda_i)M(s \mid s^{-1}\Lambda_i) = I \qquad (\Lambda_i \in (a_i)_C^*),$$

because both sides of the equation are meromorphic on the entire $(\boldsymbol{a}_i)_C^*$.

CHAPTER 6

THE GENERAL CASE OF SEVERAL CUSPS

6.1. Introduction

In this final chapter we shall carry out the whole preceding development in the general setting of several cusps. This entails a more complex combinatorial set-up of maximal cuspidal subgroups. We shall no longer be concerned with just maximal standard cuspidal subgroups. On the other hand, once the new combinatorial situation is understood, the general treatment will readily be seen to be analogous to the previous one. We will point out the generalizations and give demonstrations for the less obvious ones.

Recall that P_0 denotes the fixed minimal cuspidal subgroup. Basic to this development is the following result from reduction theory (Harish-Chandra [2, p. 5], Borel [2, 3]): The collection of double cosets $\Gamma \backslash G_Q/(P_0)_Q$ is finite. If $\Xi \subset G_Q$ contains a representative for every double coset in $\Gamma \backslash G_Q/(P_0)_Q$, then there exists a Siegel domain S_0 associate to P_0 such that $G = \Gamma \, \Xi \, S_0$. Conversely, if there exists a finite subset Ξ of G_Q and a Siegel domain S_0 associate to P_0 such that $G = \Gamma \, \Xi \, S_0$, then Ξ contains at least one representative for every double coset in $\Gamma \backslash G_Q/(P_0)_Q$.

The body of this chapter consists of five sections, corresponding to the five chapters that precede.

6.2. Definition and Basic Properties of the Eisenstein Series

As before, P_0 is the minimal cuspidal subgroup, $\{P_r\}$ is the set of maximal standard cuspidal subgroups.

Let $\Xi \subset G_Q$ be consisted of one representative from each double coset in $\Gamma \backslash G_Q/(P_0)_Q$, including the identity element e. There is a Siegel domain S_0 associate to P_0 such that $G = \Gamma\, \Xi\, S_0$. For each r, $P_0 \subset P_r$ implies

$$|\Gamma \backslash G_Q/(P_r)_Q| \leq |\Gamma \backslash G_Q/(P_0)_Q| < \infty .$$

Let $d(r) = |\Gamma \backslash G_Q/(P_r)_Q|$. Depending on r, we shall index Ξ in more than one way. Write Ξ as

$$\{\zeta_{11}, \ldots, \zeta_{1l_1}; \zeta_{21}, \ldots, \zeta_{2l_2}; \cdots; \zeta_{d(r)1}, \ldots, \zeta_{d(r)l_{d(r)}}\},$$

where $\zeta_{11} = e$, and

$$\{\zeta_{11}, \ldots, \zeta_{1l_1}\}, \ \{\zeta_{21}, \ldots, \zeta_{2l_2}\}, \ldots, \{\zeta_{d(r)1}, \ldots, \zeta_{d(r)l_{d(r)}}\}$$

are the equivalence classes of $\Gamma \backslash \Xi /(P_r)_Q$. Introduce the index sets

$$I_r = \{11, \ldots, 1l_1; 21, \ldots, 2l_2; \cdots; d(r)1, \ldots, d(r)l_{d(r)}\},$$

$$\underline{I_r} = \{11, 21, \ldots, d(r)1\}.$$

Thus $\Xi = \{\zeta_{\nu_r} : \nu_r \in I_r\}$. For $\zeta \in G$ and $P \subset G$, let ${}^{\zeta}P$ stand for $\zeta P \zeta^{-1}$.

The relevant maximal cuspidal subgroups are the following: for each r, let

$$P_{r11} = {}^{\zeta_{11}}(P_r) = P_r, \qquad P_{r21} = {}^{\zeta_{21}}(P_r), \ldots, P_{rd(r)1} = {}^{\zeta_{d(r)1}}(P_r);$$

and generally

$$P_{r\nu_r} = {}^{\zeta_{\nu_r}}(P_r) \qquad\qquad (\nu_r \in I_r).$$

Since $\{\zeta_{\nu_r} : \nu_r \in \underline{I}_r\}$ is a complete set of representatives for $\Gamma \backslash \Xi /(P_r)_Q$, there exists $\gamma_{\alpha\beta} \in \Gamma$ such that $P_{r\alpha\beta} = {}^{\gamma_{\alpha\beta}}(P_{r\alpha 1})$ $\quad (1 \le \alpha \le d(r)\, , 1 \le \beta \le l_\alpha)$.

The special case of only one cusp, i.e., $G = \Gamma S_0$, corresponds to $|\Gamma \backslash G_Q/(P_0)_Q| = 1$. So for each r we are dealing with just the maximal standard cuspidal subgroup P_r .

Write $\zeta_{\nu_r} = \zeta'_{\nu_r}\zeta''_{\nu_r}$ with $\zeta'_{\nu_r} \in K$, $\zeta''_{\nu_r} \in P_0 \subset P_r$. It is obvious that

$$P_{r\nu_r} = {}^{\zeta'_{\nu_r}}(P_r)\ .$$

Set

$$A_{r\nu_r} = {}^{\zeta'_{\nu_r}}(A_r)\ .$$

Then

$$P_{r\nu_r} = N_{r\nu_r}A_{r\nu_r}M_{r\nu_r}$$

is a Langlands' decomposition of $P_{r\nu_r}$, where

$$M_{r\nu_r} = {}^{\zeta'_{\nu_r}}(M_r)$$

and

$$N_{r\nu_r} = {}^{\zeta'_{\nu_r}}(N_r)$$

$$= {}^{\zeta_{\nu_r}}(N_r)\ .$$

As in Section 1.3, for each maximal cuspidal subgroup P_{ν_r} $(\nu_r \in \underline{I}_r\, , r \in \{1, \ldots, m\})$ Eisenstein series can be defined. These possess all the basic properties of the previous Eisenstein series; in particular by Selberg's principle they are eigenfunctions of convolution operators.

(Two or more indices written together, such as $\alpha\beta$, are to be regarded as to form a compound index, not to be confused with multiplication.)

6.3. The Compact Operators

The analysis in the general case rests on the fact that $\Gamma \backslash G$ is covered by finitely many translates of a Siegel domain: $G = \Gamma\,\Xi\,S_0$. As before, let $F \subset \Xi\,S_0$ be a fundamental domain. Recall that $P_0 = N_0A_0M_0$, $S_0 = \Omega_{N_0}A_0(t_0)K$, where $A_0(t_0) = \{a \in A_0 : \alpha_r(\log a) \leq t_0$ for each $a_r \in \Sigma^o\}$, Σ^o is the set of simple roots of $(\boldsymbol{p}_0, \boldsymbol{a}_0)$; and Ω_{N_0} is a compact subset of N_0.

As noted by Godement [3, p. 232]), his estimates of Lemma 2.4.1 are serviceable in this general situation as well.

Write Ξ as $\{\zeta_\nu\}$, where ν varies over some finite index set. The following is the general definition of a slowly increasing function.

A function f defined on G is said to be slowly increasing if there exist C, C′ > 0 such that

$$|f(\zeta_\nu g)| \leq C' e^{-C\eta(H_0(g))} \qquad \text{for } g \in S_0 \text{ and all } \nu .$$

(If we consider only automorphic functions, then it does not matter which Siegel domain S is used in this definition, as long as $G = \Gamma\,\Xi\,S$ for some finite subset Ξ of G_Q (cf. lemma 3.3.2).)

In a notation introduced in Section 1.2,

$$|{}^{\zeta_\nu}f(g)| \leq C' e^{-C\eta(H_0(g))} \qquad \text{for } g \in S_0 \text{ and a } \nu .$$

If for each ν, the arithmetic subgroup ${}^{\zeta_\nu^{-1}}\Gamma$ (cf. Borel [2]) and the function ${}^{\zeta_\nu}f$ replace Γ and f respectively, then we have the following generalization of Lemma 2.4.1. Note that

$$f \in L^2(\Gamma \backslash G) \text{ implies } {}^{\zeta_\nu}f \in L^2({}^{\zeta_\nu^{-1}}\Gamma \backslash G),$$

and that

$$\|f\|_2 = \|{}^{\zeta_\nu}f\|_2 .$$

Lemma 6.3.1. Let α be a continuous compactly supported function on G.

(i) For $f \in L^2(\Gamma \setminus G)$, there exist constants $M_\nu > 0$, which do not depend on f, such that for every $N > 0$, $r \in \{1, \ldots, m\}$, and ν, independent of f,

$$\alpha * {}^{\zeta_\nu}f(g) - \alpha * ({}^{\zeta_\nu}f)^r(g) \ll e^{-M_\nu \eta(H_0(g))} e^{N\alpha_r(H_0(g))} \|f\|_2 \qquad \text{for } g \in S_0 .$$

(ii) Let f be a Γ-automorphic continuous slowly increasing function on G, that is, there exists $C, C' > 0$ such that

$$|{}^{\zeta_\nu}f(g)| \le C' e^{-C\eta(H_0(g))} \qquad \text{for } g \in S_0 \text{ and all } \nu ,$$

then, depending on C and C', there exist constants $M_\nu > 0$ such that for every $N > 0$, $r \in \{1, \ldots, m\}$, and ν,

$$\alpha * {}^{\zeta_\nu}f(g) - \alpha * ({}^{\zeta_\nu}f)^r(g) \le B_\nu e^{-M_\nu(H_0(g))} e^{N\alpha_r(H_0(g))} \qquad \text{for } g \in S_0 ,$$

where B_ν are positive constants; the majoration is otherwise independent of f.

The functions $\alpha * {}^{\zeta_\nu}f(g)$, $\alpha * ({}^{\zeta_\nu}f)^r(g)$ in the above can be interpreted as follows. First, it is easily checked that

$$\alpha * {}^{\zeta_\nu}f(g) = \alpha * f(\zeta_\nu g) .$$

For the other functions, we introduce the maximal cuspidal subgroup $P_{r\nu}$ as defined by

$$P_{r\nu} = {}^{\zeta_\nu}(P_r) .$$

(In a different notation, namely $P_{r\nu_r}$, this set of maximal cuspidal subgroups has already been introduced in the last section. The present, less specific, notation suffices however for this and the next section.)

Let

$$P_{r\nu} = N_{r\nu} A_{r\nu} M_{r\nu}$$

be the fixed Langlands decomposition of P_{rv}, with

$$A_{rv} = {}^{\zeta'_v}(A_r)$$

and

$$N_{rv} = {}^{\zeta'_v}(N_r)$$
$$= {}^{\zeta_v}(N_r)$$

for $\zeta'_v \in K$. Denote by f^{rv} the constant term of a square integrable or continuous function f on $\Gamma \backslash G$ with respect to P_{rv}. Then

$$({}^{\zeta_v}f)^r(g) = \int_{N_r \cap {}^{\zeta_v^{-1}}\Gamma \backslash N_r} f(\zeta_v n g)\, dn = \int_{N_r \cap {}^{\zeta_v^{-1}}\Gamma \backslash N_r} f(\zeta_v n \zeta_v^{-1} \zeta_v g)\, dn$$
$$= \int_{{}^{\zeta_v}(N_r) \cap \Gamma \backslash {}^{\zeta_v}(N_r)} f(n \zeta_v g)\, dn$$

since both

$$\int_{N_r \cap {}^{\zeta_v^{-1}}\Gamma \backslash N_r} dn \quad \text{and} \quad \int_{{}^{\zeta_v}(N_r) \cap \Gamma \backslash {}^{\zeta_v}(N_r)} dn$$

are normalized to be 1. Thus

$$({}^{\zeta_v}f)^r(g) = (f^{rv})(\zeta_v g) \qquad (g \in G).$$

As in Section 2.4, *truncation factors* have to be introduced before we can utilize these estimates. Of course, the combinatorics of the present situation is more complex. Let e_{rv} be the continuous functions on G defined by $e_{rv} = {}^{\zeta_v^{-1}}(\delta_r)$, where δ_r are the truncation factors introduced in Section 2.4. Let $S = \tilde{\Omega}_N A(t) K$ be another Siegel domain, sufficiently large so that its interior S^o contains S_0. Then $G = \Gamma\, \Xi\, S_0 = \Gamma\, \Xi\, S^o$. Clearly $\Xi\, S^o$ is an open submanifold of G. Let $\{f_\mu\}$ be a partition of unity subordinate to the open cover $\{\zeta_v S^o\}$ of $\Xi\, S^o$.

Denote by J_ν the subset $\{\mu : \text{supp}\, f_\mu \subset \zeta_\nu S^o\}$ of the index set $\{\mu\}$. Let

$$\omega_\nu'(g) = \sum_{\mu \in J_\nu} f_\mu(g) \qquad (g \in \Xi S^o).$$

It is clear that ω_ν' are well defined, continuous functions on ΞS^o. Moreover, $\omega_\nu'(g) = 0$ for $g \notin \zeta_\nu S^o$. Now let

$$\omega_\nu = \frac{\omega_\nu'}{\sum_\lambda \omega_\lambda'}$$

(λ varies over the same finite index set as ν). Since $\bigcup_\lambda J_\lambda = \{\mu\}$, $\sum_\lambda \omega_\lambda'(g) \geq 1 > 0$ for all $g \in \Xi S^o$. So ω_ν are well defined, continuous functions on ΞS^o. By virtue of their construction, they have the following properties: $0 \leq \omega_\nu(g) \leq 1$, $\omega_\nu(g) = 0$ for $g \notin \zeta_\nu S^o$, and

$$\sum_\nu \omega_\nu(g) = 1.$$

Lastly, set

$$\delta_{r\nu}(g) = \omega_\nu(g) e_{r\nu}(g) \qquad \text{for } g \in \Xi S^o.$$

The truncation factors $\delta_{r\nu}$ are continuous functions on ΞS^o,

$$0 \leq \delta_{r\nu}(g) \leq 1, \quad \delta_{r\nu}(g) = 0 \qquad \text{for } g \notin \zeta_\nu S^o,$$

and

$$\sum_{r\nu} \delta_{r\nu}(g) = \sum_{r=1}^{m} \sum_\nu \delta_{r\nu}(g) = \sum_\nu \omega_\nu(g) \sum_{r=1}^{m} e_{r\nu}(g)$$

$$= \sum_\nu \omega_\nu(g) \sum_{r=1}^{m} \delta_r(\zeta_\nu^{-1} g) = \sum_\nu \omega_\nu(g) = 1 \qquad \text{for all } g \in \Xi S^o.$$

The corresponding generalization of Lemma 2.4.2 is as follows, the proof of which is analogous to that of Lemma 2.4.2.

Lemma 6.3.2. Let α be a continuous compactly supported function on G.

(i) There exists $\kappa > 0$ such that for all $f \in L^2(\Gamma \setminus G)$,

$$|\alpha * f(g) - \sum_{rv} \delta_{rv}(g)\alpha * f^{rv}(g)| \leq \kappa \|f\|_2 \qquad \text{for } g \in \Xi S_0 .$$

(ii) Let f be a Γ-automorphic continuous slowly increasing function on G, that is, there exists $C, C' > 0$ such that

$$|f(\zeta_v g)| \leq C' e^{-C\eta(H_0(g))} \qquad \text{for } g \in S_0 \text{ and all } v .$$

Then, depending on C and C', there exists $\kappa > 0$ such that

$$|\alpha * f(g) - \sum_{rv} \delta_{rv}(g)\alpha * f^{rv}(g)| \leq \kappa \qquad \text{for } g \in \Xi S_0 ;$$

κ is otherwise independent of f.

Proof. We give only the proof of (i). The proof of (ii) is analogous.

Lemma 6.3.1 will be applied with S in place of S_0. Since

$$\sum_{rv} \delta_{rv}(g) = 1 \qquad (g \in \Xi S^o \supset \Xi S_0),$$

we have

$$|\alpha * f(g) - \sum_{rv} \delta_{rv}(g)\alpha * f^{rv}(g)| = |\sum_{r=1}^{m} \sum_{v} \delta_{rv}(g)(\alpha * f(g) - \alpha * f^{rv}(g))| .$$

Fix $r \in \{1, \ldots, m\}$ and consider

$$|\sum_{v} \delta_{rv}(g)(\alpha * f(g) - \alpha * f^{rv}(g))| \qquad \text{for } g \in \Xi S^o \supset \Xi S_0 .$$

As $\delta_{r\nu}(g) = 0$ for $g \notin \zeta_\nu S^o$, we can assume that $g \in \zeta_\nu S^o$. If we write $g = \zeta_\nu g'$, where $g' \in S^o$, then $e_{r\nu}(g) = \delta_r(\zeta_\nu^{-1} g) = \delta_r(g')$. So we consider

$$|\sum_\nu \omega_\nu(\zeta_\nu g')\delta_r(g')(\alpha * f(\zeta_\nu g') - \alpha * f^{r\nu}(\zeta_\nu g'))| \qquad \text{for } g' \in S^o \subset S.$$

In the notation of the proof of Lemma 2.4.2, either $g' \in S^{(r)}$, or $g' \in S^{(p)}$ for some $p \neq r$. In Lemma 6.3.1(i), choose $N > 0$ such that $N - M_\nu > 0$ for all ν. Suppose $g' \in S^{(r)}$. Then $\eta(H_0(g')) = \alpha_r(H_0(g'))$. Therefore,

$$e^{-M_\nu \eta(H_0(g'))} e^{N\alpha_r(H_0(g'))} = e^{(N-M_\nu)\alpha_r(H_0(g'))}$$

$$\leq e^{(N-M_\nu)t},$$

because $g' \in S$. Thus Lemma 6.3.1(i), together with the remark that followed it, gives for every ν that

$$|\alpha * f(\zeta_\nu g') - \alpha * f^{r\nu}(\zeta_\nu g')| \leq B_\nu e^{(N-M_\nu)t} \|f\|_2 \qquad \text{for some } B_\nu > 0.$$

Then

$$|\sum_\nu \omega_\nu(\zeta_\nu g')\delta_r(g')(\alpha * f(\zeta_\nu g') - \alpha * f^{r\nu}(\zeta_\nu g'))| \leq B\|f\|_2 \qquad \text{for some } B > 0.$$

On the other hand, if instead $g' \in S^{(p)}$ for some $p \neq r$, then, as in the proof of Lemma 2.4.2,

$$|\alpha * f(\zeta_\nu g') - \alpha * f^{r\nu}(\zeta_\nu g')| \leq C_\nu e^{(N-M_\nu)t+N} \|f\|_2 \qquad \text{for some } C_\nu > 0,$$

so that

$$|\sum_\nu \omega_\nu(\zeta_\nu g')\delta_r(g')(\alpha * f(\zeta_\nu g') - \alpha * f^{r\nu}(\zeta_\nu g'))| \leq D\|f\|_2 \qquad \text{for some } D > 0.$$

It is now clear that there exists $\kappa > 0$ such that

$$|\sum_{r=1}^{m}\sum_\nu \delta_{r\nu}(g)(\alpha * f(g) - \alpha * f^{r\nu}(g))| \leq \kappa\|f\|_2 \qquad \text{for } g \in \Xi S_0. \quad \square$$

We now present the definition of the compact operators for the analytic continuation of the Eisenstein series in the general situation. Let α be a smooth compactly supported function on G. Define an operator $\mathbf{K}_\alpha$ on $L^2(\Gamma \setminus G)$ as follows:

For $f \in L^2(\Gamma \setminus G)$, let

$$\mathbf{K}_\alpha^o f = \left[\alpha * f - \sum_{rv} \delta_{rv} \alpha * f^{rv} \right]_F ,$$

and

$$\mathbf{K}_\alpha = \lambda_\alpha \mathbf{K}_\alpha^o \qquad \text{where } \lambda_\alpha = \alpha * \alpha .$$

It is easily checked that Theorem 2.4.1, Proposition 2.4.1, and Proposition 2.4.2 hold in general.

6.4. Fredholm Equations

In this and the next section, we fix a maximal standard cuspidal subgroup, say P_1, and $\zeta_{\alpha_0 1} \in \Xi$ $(\alpha_0 1 \in \underline{I_1})$, and for simplicity let P stand for $P_{1\alpha_0 1} = {}^{\zeta_{\alpha_0 1}}(P_1)$. We consider a cuspidal Eistenstein series $E(\Lambda, \Phi, g)$ for the maximal cuspidal subgroup P.

The compact operator defined in the last section involves every constant term with respect to a maximal cuspidal subgroup P_{rv}. In Section 3.2, Lemma 3.2.1 and Theorem 3.2.1 needed not be restricted to standard cuspidal subgroups.

Lemma 6.4.1. Suppose P_{rv} is a maximal cuspidal subgroup not associate to P, then

$$E^{rv}(\Lambda, g) = 0 \qquad ((\Lambda, g) \in (\mathfrak{a}_C^*)^+ \times G) .$$

Note that P_{rv} is associated with P if and only if P_r is associate to P. Hence, using the notation of Section 3.2, among the relevant maximal cuspidal subgroups, $P_{j\nu}$ are those which are associate to P .

Theorem 6.4.1. Suppose $P_{j\nu}$ is a maximal cuspidal subgroup associate to P , then

$$E^{j\nu}(\Lambda, g) = \sum_{s \in W(\boldsymbol{a}, \boldsymbol{a}_{j\nu})} e^{(-s\Lambda-\rho_{j\nu})(\log a_{j\nu})}\sigma(k)(M(s \mid \Lambda)\Phi)(m_{j\nu})$$

($\Lambda \in (\boldsymbol{a}_C^*)^+$; $g = n_{j\nu}a_{j\nu}m_{j\nu}k$, $n_{j\nu} \in N_{j\nu}$, $a_{j\nu} \in A_{j\nu}$, $m_{j\nu} \in M_{j\nu}$, $k \in K$), where $W(\boldsymbol{a}, \boldsymbol{a}_{j\nu})$ is the Weyl group of $(\boldsymbol{a}, \boldsymbol{a}_{j\nu})$; $M(s \mid \Lambda)$ is a linear transformation from ${}^oL^2(\Gamma_M \backslash M, \sigma_M, \chi)$ into ${}^oL^2(\Gamma_{M_{j\nu}} \backslash M_{j\nu}, \sigma_{M_{j\nu}}, {}^s\chi)$.

As in Section 3.2, we rewrite Langlands' formula as:

$$E^{j\nu}(\Lambda, g) = \sum_{s,n} \phi(j\nu, s, n \mid \Lambda)e^{(-s\Lambda-\rho_{j\nu})(\log a_{j\nu})}\sigma(k)\Psi_{j\nu,s,n}(m_{j\nu}) ,$$

where the constant term coefficients $\phi(j\nu, s, n \mid \Lambda)$ are defined and holomorphic on $(\boldsymbol{a}_C^*)^+$.

For every $(j\nu, s, n)$, let

$$I_{j\nu,s,n}(\Lambda, g) = \delta_{j\nu}(g)e^{(-s\Lambda-\rho_{j\nu})(\log a_{j\nu})}\sigma(k)\Psi_{j\nu,s,n}(m_{j\nu}) \qquad ((\Lambda, g) \in \boldsymbol{a}_C^* \times \Xi S^o) ,$$

and define a function $H^o_{j\nu,s,n}(\Lambda, g)$ on $\boldsymbol{a}_C^* \times G$ by

$$H^o_{j\nu,s,n} = \lambda_\alpha[\, I_{j\nu,s,n} \,]_F \qquad \text{for each } \Lambda .$$

In section 3.4 we occupied ourselves with proving that $[\, I_{j,s,n} \,]_F$ is slowly increasing. Now we sketch the proof for the general case of $[\, I_{j\nu,s,n} \,]_F$. Then Proposition 3.3.4 and Lemma 6.3.2(ii) will imply the projection of constant terms (cf. Theorem 3.4.2).

First of all, Proposition 6.4.1 takes the place of Proposition 3.4.1.

Proposition 6.4.1. There exist real-valued continuous functions $d_r(\Lambda)$ $(r = 1, \ldots, m)$ on $\boldsymbol{a}_C^*$ and $C > 0$ such that

$$|I_{j\nu,s,n}(\Lambda, g)| \leq C \exp\left[\sum_{r=1}^{m} d_r(\Lambda)\alpha_r(H_0(\zeta_\nu'^{-1}g))\right] \qquad ((\Lambda, g) \in \boldsymbol{a}_C^* \times \Xi S^o).$$

Proof. As usual $H_{j\nu}(g) = \log a_{j\nu}$ for $g = n_{j\nu}a_{j\nu}m_{j\nu}k$. Then

$$I_{j\nu,s,n}(\Lambda, g) = \delta_{j\nu}(g)e^{(-s\Lambda-\rho)(H_{j\nu}(g))}\sigma(k)\Psi_{j\nu,s,n}(m_{j\nu}).$$

Let us consider the exponential $e^{(-s\Lambda-\rho_{j\nu})(H_{j\nu}(g))}$. Recall that $A_{j\nu} = {}^{\zeta_\nu'}(A_j)$, where $\zeta_\nu' \in K$. Let $\tilde{s}$ be the element of $W(a_{j\nu}, a_j)$ corresponding to $\zeta_\nu'^{-1}$. Then

$$e^{(-s\Lambda-\rho_{j\nu})(H_{j\nu}(g))} = e^{(-\tilde{s}s\Lambda-\tilde{s}\rho_{j\nu})({}^{\zeta_\nu'^{-1}}H_{j\nu}(g))}.$$

For $g = n_ja_jm_jk$, where $n_j \in N_j$, $a_j \in A_j$, $m_j \in M_j$, $k \in K$,

$${}^{\zeta_\nu'}g = {}^{\zeta_\nu'}(n_j){}^{\zeta_\nu'}(a_j){}^{\zeta_\nu'}(m_j){}^{\zeta_\nu'}k,$$

where

$${}^{\zeta_\nu'}(n_j) \in N_{j\nu}, \quad {}^{\zeta_\nu'}(a_j) \in A_{j\nu}, \quad {}^{\zeta_\nu'}(m_j) \in M_{j\nu}, \quad {}^{\zeta_\nu'}k \in K.$$

It is clear that

$${}^{\zeta_\nu'^{-1}}H_{j\nu}({}^{\zeta_\nu'}g) = \log{}^{\zeta_\nu'^{-1}}({}^{\zeta_\nu'}(a_j)) = \log a_j = H_j(g) \quad \text{for } g \in G,$$

which implies

$${}^{\zeta_\nu'^{-1}}H_{j\nu}(g) = H_j({}^{\zeta_\nu'^{-1}}g) = H_j(\zeta_\nu'^{-1}g).$$

Referring to the proof of Proposition 3.4.1, we know that there exist real-valued continuous functions $d_r(\Lambda)$ and $C > 0$ such that

$$|I_{j\nu,s,n}(\Lambda, g)| \leq C \exp\left[\sum_{r=1}^{m} d_r(\Lambda)\alpha_r(H_0(\zeta_\nu'^{-1}g))\right] \qquad ((\Lambda, g) \in \boldsymbol{a}_C^* \times \Xi S^o). \qquad \square$$

Fix a compact subset E of $\mathfrak{a}_C^*$. Suppose $\Lambda \in E$, $g \in S_0$, $\zeta_\lambda \in \Xi$, and there exist $g' \in S_0$, $\zeta_{\lambda'} \in \Xi$, $\zeta_{\lambda'}g' \in F \subset \Xi S_0$, and $\gamma \in \Gamma$ such that $\gamma\zeta_{\lambda'}g' = \zeta_\lambda g$. Then, by the very definition of $[\, I_{j\nu,s,n}\,]_F$,

$$[\, I_{j\nu,s,n}\,]_F(\Lambda, \zeta_\lambda g) = I_{j\nu,s,n}(\Lambda, \zeta_{\lambda'}g')$$

$$= I_{j\nu,s,n}(\Lambda, \gamma^{-1}\zeta_\lambda g)\,,$$

and Proposition 6.4.1 implies

$$|\, [\, I_{j\nu,s,n}\,]_F(\Lambda, \zeta_\lambda g)\, | \;\le\; C\exp\left[\, \sum_{r=1}^{m} d_r(\Lambda)\alpha_r(H_0(\zeta_\nu'^{-1}\gamma^{-1}\zeta_\lambda g))\right].$$

To relate $\alpha_r(H_0(\zeta_\nu'^{-1}\gamma^{-1}\zeta_\lambda g))$ $(r = 1, \ldots, m)$ to $\eta(H_0(g))$ for $g\ \Xi\ S_0$ and $\gamma^{-1} \in \Gamma$ such that $\gamma^{-1}\zeta_\lambda g \in \Xi S_0$, recall from reduction theory that $\{\, \gamma \in \Gamma : \gamma\,\Xi S_0 \cap \Xi S_0 \neq \phi\, \}$ is a finite set. This is the Siegel property of Siegel domains. Hence $\zeta_\nu'^{-1}\gamma^{-1}\zeta_\lambda$ belongs to a certain finite set $\{\omega_k\}$. The rest of the argument proceeds along the same line as in Section 3.4. This leads eventually to that $[\, I_{j\nu,s,n}\,]_F$ is slowly increasing, uniformly over the compact subset E.

To simplify the notations, we shall often use a multiindex $\bar{J}$ in place of $(j\nu, s, n)$.

All the statements in Sections 3.5 and 3.6 remain valid in the present general situation. In particular, we have

$$\mathbf{K}_\alpha E(\Lambda, g) = \hat{\alpha}(\Lambda)E(\Lambda, g) - \sum_{j\nu} \hat{\alpha}^o(\Lambda)\lambda_\alpha[\, \delta_{j\nu}E^{j\nu}\,]_F(\Lambda, g) \qquad (\Lambda \in (\mathfrak{a}_C^*)^+)\,;$$

If f is Γ-automorphic, slowly increasing, $\mathbf{K}$-finite and $\mathbf{Z}(G)$-finite, then

$$f - \sum_{r\nu} [\, \delta_{r\nu}f^{r\nu}\,]_F$$

is bounded; the Fredholm equations are

$$(\, \mathbf{K}_\alpha - \hat{\alpha}(\Lambda)\,)^o F_{\bar{J}}^{**}(\Lambda) = -\, \mathbf{K}_\alpha H_{\bar{J}}^o(\Lambda) \qquad (\, \Lambda \in \mathfrak{a}_C^* \setminus \hat{\alpha}^{-1}(\operatorname{spec} \mathbf{K}_\alpha)\,) \qquad \text{for all } \bar{J}\,;$$

and the Eisenstein series

$$E(\Lambda) = \sum_{\bar{J}} \phi(\bar{J} \mid \Lambda) F^*_{\bar{J}}(\Lambda) = \sum_{j\nu,s,n} \phi(j\nu,\, s,\, n \mid \Lambda) F^*_{j\nu,s,n}(\Lambda) \qquad (\, \Lambda \in (\boldsymbol{a}^*_C)^+ \setminus \hat{\alpha}^{-1}(\text{spec } \mathbf{K}_\alpha)\,)\,.$$

6.5. Analytic Continuation

From the last section

$$E(\Lambda, g) = \sum_{\bar{J}} \phi(\bar{J} \mid \Lambda) F^*_{\bar{J}}(\Lambda, g) = \sum_{j\nu,\, s,\, n} \phi(j\nu,\, s,\, n \mid \Lambda) F^*_{j\nu,s,n}(n, g) \qquad (g \in G)$$

for $\Lambda \in (\boldsymbol{a}^*_C)^+ \setminus \hat{\alpha}^{-1}(\text{spec } \mathbf{K}_\alpha)$. As in Section 4.2, we set up a system of linear equations in $\phi(\bar{J} \mid \Lambda)$:

$$(6.5.1') \qquad \sum_{\bar{J}} \phi(\bar{J} \mid \Lambda)[\, DF^*_{\bar{J}}(\Lambda, g) - \chi_\Lambda(D) F^*_{\bar{J}}(\Lambda, g)\,] = 0 \qquad (g \in G),\ D \in \mathbf{Z}(G)\,;$$

if P_i is a maximal standard cuspidal subgroup associate to P, and ${}^{\zeta_\lambda}(P_i) = P_{i\lambda} = N_{i\lambda} A_{i\lambda} M_{i\lambda}$, then for $g = n_{i\lambda} a_{i\lambda} m_{i\lambda} k$ where $n_{i\lambda} \in N_{i\lambda}$, $a_{i\lambda} \in A_{i\lambda}$, $m_{i\lambda} \in M_{i\lambda}$, $k \in K$,

$$(6.5.2') \qquad \sum_{j\nu,s,n} \phi_{j\nu,s,n}(\Lambda) \left[\int_{N_{i\lambda} \cap \Gamma \setminus N_{i\lambda}} F^*_{j\nu,s,n}(\Lambda, ng)\, dn - \delta_{j\nu,i\lambda} e^{(-s\Lambda - \rho_{i\lambda})(\log a_{i\lambda})} \sigma(k) \Psi_{i\lambda,s,n}(m_{i\lambda}) \right] = 0\,,$$

$i\lambda$ *varies over the same index as* $j\nu$ ($\delta_{j\nu,i\lambda} = 1$ if $j = i$ and $\nu = \lambda$, 0 otherwise); if P_l is a maximal standard cuspidal subgroup not associate to P, then

$$(6.5.3') \qquad \sum_{\bar{J}} \phi(\bar{J} \mid \Lambda) \int_{N_l \cap \Gamma \setminus N_l} F^*_{\bar{J}}(\Lambda, ng)\, dn = 0 \qquad (g \in G),\ l \in L\,;$$

$$(6.5.4') \qquad \sum_{\bar{J}} \phi(\bar{J} \mid \Lambda)[\, F^*_{\bar{J}}(\Lambda, gk) - \sigma(k) F^*_{\bar{J}}(\Lambda, g)\,] = 0 \qquad (g \in G),\ k \in K\,.$$

We are here dealing with an Eisenstein series for $P = P_{1\alpha_0 1}$. In the notations of Sections 6.2 and 6.3, for $1 \le \beta \le l_{\alpha_0}$,

$$\zeta'_{\alpha_0\beta}\zeta'^{-1}_{\alpha_0 1}(P, A) = \left(\zeta'_{\alpha_0\beta}\zeta'^{-1}_{\alpha_0 1}P,\ \zeta'_{\alpha_0\beta}\zeta'^{-1}_{\alpha_0 1}A\right) = (P_{1\alpha_0\beta}, A_{1\alpha_0\beta}) .$$

For simplicity, let ν' denote $\alpha_0\beta$, and write $(P_{\nu'}, A_{\nu'})$ for $(P_{1\alpha_0\beta}, A_{1\alpha_0\beta})$; moreover, let $s_{\nu'}$ denote the elements of $W(\boldsymbol{a}, \boldsymbol{a}_{\nu'})$ corresponding to $\zeta'_{\alpha_0\beta}\zeta'^{-1}_{\alpha_0 1}$ $(1 \le \beta \le l_{\alpha_0})$.

For simplicity, we shall often use a multiindex $\bar{J}$ in place of $(1\nu', s_{\nu'}, n)$ (i.e., $j = 1$, $\nu = \nu'$, $s = s_{\nu'}$). We will show that $\phi(\bar{J}_1 \mid \Lambda)$ can easily be extended to entire functions on $\boldsymbol{a}_C^*$. First we state a proposition, whose proof requires only straightforward verification.

Proposition 6.5.1. Let f be a square integrable or continuous function on $\Gamma \setminus G$. If P_1 and P_2 are any two Γ-conjugate cuspidal subgroups, i.e., ${}^{\gamma}P_1 = P_2$ for some $\gamma \in \Gamma$, then $f^1(g) = f^2(\gamma g)$.

For simplicity, write $E^{\alpha_0 1}$ for $E^{1\alpha_0 1}$, $E^{\nu'}$ for $E^{1\nu'}$. Now

$$E^{\alpha_0 1}(\Lambda, g) = \sum_{s \in W(\boldsymbol{a}, \boldsymbol{a})} e^{(-s\Lambda-\rho)(H(g))}\sigma(k)(M(s \mid \Lambda)\Phi)(m)$$

$(\Lambda \in (\boldsymbol{a}_C^*)^+$; $g = namk$, $n \in N$, $a \in A$, $m \in M$, $k \in K)$. As shown in Section 4.2, $M(1 \mid \Lambda)$ is the identity transformation on ${}^0L^2(\Gamma_M \setminus M, \sigma_M, \chi)$, and this implies that $\phi_{1\alpha_0 1,1,n}(\Lambda)$ are actually constant. More generally

$$E^{\nu'}(\Lambda, g) = \sum_{s \in W(\boldsymbol{a}, \boldsymbol{a}_{\nu'})} e^{(-s\Lambda-\rho_{\nu'})(H_{\nu'}(g))}\sigma(k)(M(s \mid \Lambda)\Phi)(m_{\nu'})$$

$(\Lambda \in (\boldsymbol{a}_C^*)^+$; $g = n_{\nu'}a_{\nu'}m_{\nu'}k$, $n_{\nu'} \in N_{\nu'}$, $a_{\nu'} \in A_{\nu'}$, $m_{\nu'} \in M_{\nu'}$, $k \in K)$. Given ν', there exists $\gamma_{\nu'} \in \Gamma$ such that

$${}^{\gamma_{\nu'}}P = P_{\nu'} ,$$

hence

$$E^{\alpha_0 1}(g) = E^{\nu'}(\gamma_{\nu'} g)$$

by Proposition 6.5.1. Let us find the decomposition of $\gamma_{v'}g$ in $N_{v'}A_{v'}M_{v'}K$. For simplicity, let ζ' denote $\zeta'_{\alpha_0\beta}\zeta'^{-1}_{\alpha_0 1} \in K$. We have also

$$^{\zeta'}P = P_{v'}.$$

Then $\zeta'^{-1}\gamma_{v'} = n'a'm'$ where $n' \in N$, $a' \in A$, $m' \in M$, which implies that $\gamma_{v'} = \zeta' n'a'm'$. If now $g = namk$, where $n \in N$, $a \in A$, $m \in M$, $k \in K$, then

$$\begin{aligned}\gamma_{v'}g &= \zeta' n'a'm'knamk \\ &= \zeta' n'(a'm')n(a'm')^{-1}a'm'amk = \zeta' n'(a'm')n(a'm')^{-1}a'am'mk \\ &= \zeta'[n'(a'm')n(a'm')^{-1}]\zeta'^{-1}\zeta'(a'a)\zeta'^{-1}\zeta'(m'm)\zeta'^{-1}\zeta'k,\end{aligned}$$

where

$$\zeta'[n'(a'm'n(a'm')^{-1}]\zeta'^{-1} \in N_{v'}, \quad \zeta'(a'a)\zeta'^{-1} \in A_{v'}, \quad \zeta'(m'm)\zeta'^{-1} \in M_{v'}, \quad \zeta'k \in K.$$

Thus

$$H_{v'}(\gamma_{v'}g) = \log \zeta'(a'a)\zeta'^{-1} = \log\zeta'a'\zeta'^{-1} + \log \zeta'a\zeta'^{-1},$$

and

$$\begin{aligned}(6.5.5)\quad E^{v'}(\Lambda, \gamma_{v'}g) &= e^{(-s_{v'}\Lambda-\rho_{v'})(\log \zeta'a\zeta'^{-1})}e^{(-s_{v'}\Lambda-\rho_{v'})(\log \zeta'a'\zeta'^{-1})} \\ &\quad\times \sigma(\zeta'k)(M(s_{v'} \mid \Lambda)\Phi)(\zeta'(m'm)\zeta'^{-1}) \\ &\quad+ \sum_{s \in W(\mathfrak{a}, \mathfrak{a}_{v'}) \setminus \{s_{v'}\}} e^{(-s\Lambda-\rho_{v'})(\log \zeta'a'\zeta'^{-1})}e^{(-s\Lambda-\rho_{v'})(\log \zeta'a'\zeta'^{-1})} \\ &\quad\times \sigma(\zeta'k)(M(s \mid \Lambda)\Phi)(\zeta'(m'm)\zeta'^{-1})\end{aligned}$$

On the other hand,

$$\begin{aligned}(6.5.6)\quad E^{\alpha_0 1}(\Lambda, g) &= \sum_{s \in W(\mathfrak{a}, \mathfrak{a})} e^{(-s\Lambda-\rho)(\log a)}\sigma(k)(M(s \mid \Lambda)\Phi)(m) \\ &= \sum_{s \in W(\mathfrak{a}, \mathfrak{a})} e^{(-s_{v'}s\Lambda-s_{v'}\rho)(\log \zeta'a\zeta'^{-1})}\sigma(k)(M(s \mid \Lambda)\Phi)(m)\end{aligned}$$

$$= e^{(-s_{v'}\Lambda - s_{v'}\rho)(\log \zeta' a \zeta'^{-1})} \sigma(k)\Phi(m)$$

$$+ \sum_{s \in W(\boldsymbol{a}, \boldsymbol{a}) \setminus \{1\}} e^{(-s_{v'}s\Lambda - s_{v'}\rho)(\log \zeta' a \zeta'^{-1})} \sigma(k)(M(s \mid \Lambda)\Phi)(m) .$$

Note that

$$W(\boldsymbol{a}, \boldsymbol{a}_{v'}) = s_{v'} W(\boldsymbol{a}, \boldsymbol{a}) .$$

Also, $s_{v'}\rho = \rho_{v'}$ because ${}^{\zeta'}(P, A) = (P_{v'}, A_{v'})$, ζ' corresponding to $s_{v'}$. By comparing terms in equations (6.5.5) and (6.5.6), we have, in particular, that

$$e^{(-s_{v'}\Lambda - \rho_{v'})(\log \zeta' a' \zeta'^{-1})} \sigma(\zeta' k)(M(s_{v'} \mid \Lambda)\Phi)(\zeta'(m'm)\zeta'^{-1}) = \sigma(k)\Phi(m)$$

($\Lambda \in (\boldsymbol{a}_C^*)^+$; $g = namk$, $n \in N$, $a \in A$, $k \in K$). Since this identity holds for all $\Phi \in {}^oL^2(\Gamma_M \setminus M, \sigma_M, \chi)$, $M(s_{v'} \mid \Lambda)$ can be extended to entire functions on $\boldsymbol{a}_C^*$, and then $\phi(\overline{J}_1 \mid \Lambda) = \phi(1v', s_{v'}, n \mid \Lambda)$ are holomorphic on $\boldsymbol{a}_C^*$ for any choice of Φ .

We now rewrite equations (6.5.1′), (6.5.2′), (6.5.3′), and (6.5.4′) as was done in Section 4.2. Introduce the following modification of jv , s , and $\overline{J}$: for each j let I_j be the index set introduced in Section 6.2, $\overline{J}' = (jv'', s', n)$ where

$$v'' \in \begin{cases} I_1 \setminus \{v'\} & \text{if } j = 1 \\ I_j & \text{if } j \neq 1 , \end{cases}$$

$$s' \in \begin{cases} W(\boldsymbol{a}, \boldsymbol{a}_{v'}) \setminus \{s_{v'}\} & \text{if } j = 1, v = v' \ \ (v' = \alpha_0\beta,\ 1 \le \beta \le l_{\alpha_0}) \\ W(\boldsymbol{a}, \boldsymbol{a}_{v''}) & \text{otherwise} . \end{cases}$$

Fix $\Lambda \in (\boldsymbol{a}_C^*)^+ \setminus \hat{\alpha}^{-1}(\operatorname{spec}\mathbf{K}_\alpha)$. We have the following system of linear equations with $A_{D,\overline{J}'}(g)$, $B_{i\lambda,\overline{J}'}(g)$, $C_{l,\overline{J}'}$, $D_{k,\overline{J}'}(g)$ as coefficients and $\phi(\overline{J}')$ as unknowns:

$$\sum_{\overline{J}'} \phi(\overline{J}')A_{D,\overline{J}'}(g) = -\sum_{\overline{J}_1} \phi(\overline{J}_1)A_{D,\overline{J}_1}(g), \qquad D \in \mathbf{Z}(G) ,$$

$$\sum_{\bar{J}'} \phi(\bar{J}')B_{i\lambda,\bar{J}'}(g) = -\sum_{\bar{J}_1} \phi(\bar{J}_1)B_{i\lambda,\bar{J}_1}(g),$$

$i\lambda$ *varies over the same index set as* $j\nu$,

$$\sum_{\bar{J}'} \phi(\bar{J}')C_{l,\bar{J}'}(g) = -\sum_{\bar{J}_1} \phi(\bar{J}_1)C_{l,\bar{J}_1}(g), \qquad l \in L ,$$

$$\sum_{\bar{J}'} \phi(\bar{J}')D_{k,\bar{J}'}(g) = -\sum_{\bar{J}_1} \phi(\bar{J}_1)D_{k,\bar{J}_1}(g), \qquad k \in K,$$

$$g \in G .$$

Denote the cardinality of $\{\phi(\bar{J}')\}$ by m . If the determinant of every m equation from the linear system vanishes, then there exists a *nonzero* function $F^-(g)$ given by $\sum_{\bar{J}} \phi_{\bar{J}}^{\,-} F_{\bar{J}}^*(g)$ which possesses properties similar to those of $F(g) = \sum_{\bar{J}} \phi_{\bar{J}} F_{\bar{J}}^*(g)$, or the Eisenstein series; the crucial difference between $F(g)$ and $F^-(g)$ is that $\phi_{\bar{J}_1}^{\,-} = 0$ for the latter. However, for appropriately restricted $\Lambda \in (\boldsymbol{a}_C^*)^+ \setminus \hat{\alpha}^{-1}(\text{spec } \mathbf{K}_\alpha)$, this cannot happen. Consequently, with such a choice of Λ , $\phi(\bar{J}' \mid \Lambda)$ can be expressed in terms of $A_{D,\bar{J}}(\Lambda, g)$, $B_{i\lambda,\bar{J}}(\Lambda, g)$, $C_{l,\bar{J}}(\Lambda, g)$, $D_{k,\bar{J}}(g)$, $\phi(\bar{J}_1 \mid \Lambda)$ $(g \in G)$, and the generalization of Theorem 4.3.2 can be deduced. This development is much like that of Section 4.3; we elaborate only on the generalization of Lemma 4.3.6.

For $\lambda_1'' \in I_1 \setminus \{\alpha_0 1, \ldots, \alpha_0 l_{\alpha_0}\}$, $P_{1\lambda_1''}$ is G_Q-conjugate, but not Γ-conjugate, to P . For simplicity, write $P_{\lambda_1''}$ for $P_{1\lambda_1''}$. Let $s_{\lambda_1''}$ be the element of $W(\boldsymbol{a}, \boldsymbol{a}_{\lambda_1''})$ corresponding to $\zeta'_{\lambda_1''}\zeta'^{-1}_{\alpha_0 1}$.

<u>Proposition 6.5.2.</u> In the equation

$$E^{\lambda_1''}(\Lambda, g) = \sum_{N_{\lambda_1''} \cap \Gamma \setminus N_{\lambda_1''}} E(\Lambda, ng)\, dn$$

$$= \sum_{s \in W(\boldsymbol{a}, \boldsymbol{a}_{\lambda_1''})} e^{(-s\Lambda - \rho_{\lambda_1''})(\log a_{\lambda_1''})} (M(s \mid \Lambda)\Phi)(m_{\lambda_1''}) ,$$

$M(s_{\lambda_1''} \mid \Lambda) = 0$.

Proof. Recall that

$$\int_{N_{\lambda_1''} \cap \Gamma \backslash N_{\lambda_1''}} \int_{\Gamma \cap P \backslash \Gamma(s_{\lambda_1''})} \Phi_\Lambda(\gamma n a_{\lambda_1''} m_{\lambda_1''})\, dn$$

$$= e^{(-s_{\lambda_1''}\Gamma - \rho_{\lambda_1''})(\log a_{\lambda_1''})} (M(s_{\lambda_1''} \mid \Lambda)\Phi)(m_{\lambda_1''}) ,$$

where

$$\Gamma(s_{\lambda_1''}) = \Gamma \cap P_{\lambda_1''}(\zeta'_{\lambda_1''} \zeta'^{-1}_{\alpha_0 1})P .$$

(For simplicity, we have written n for $n_{\lambda_1''}$.) We claim that $\Gamma(s_{\lambda_1''})$ is empty. Let ζ' denote $\zeta'_{\lambda_1''} \zeta'^{-1}_{\alpha_0 1}$. Suppose

$$\gamma = q\zeta' p \in \Gamma \cap (P_{\lambda_1''})\zeta' P ,$$

then

$$^{\zeta'}P = P_{\lambda_1''} \quad \text{implies} \quad {}^{(q\zeta' p)}P = P_{\lambda_1''} ,$$

i.e., $^{\gamma}P = P_{\lambda_1''}$, which contradicts that P and $P_{\lambda_1''}$ are not Γ-conjugates. □

Let us now recall the important Lemma 4.3.4.

(Lemma 4.3.4.) *The number of maximal standard cuspidal subgroups associate to* P_1 *, a given maximal standard cuspidal subgroup, is either* 1 *or* 2 *. In the first case*

$$W(\boldsymbol{a}_1, \boldsymbol{a}_1) = \{1, s\} ,$$

where $sH_1 = -H_1$ *for all* $H_1 \in \boldsymbol{a}_1$ *. In the second case, suppose* P_2 *is the other maximal standard cuspidal subgroup. If* α_1 *, respectively* α_2 *, is the simple root of* (P_1, A_1) *, respectively* (P_2, A_2) *, and* $H_1 \in \boldsymbol{a}_1$ *,* $H_2 \in \boldsymbol{a}_2$ *are such that* $\alpha_1(H_1) = 1$ *,* $\alpha_2(H_2) = 1$ *, then*

$$W(\boldsymbol{a}_1, \boldsymbol{a}_2) = \{s\} ,$$

where $sH_1 = -H_2$.

For $\lambda_1 \in I_1$, let $s_{1\lambda_1}$ be the element of $W(\boldsymbol{a}_1, \boldsymbol{a}_{1\lambda_1})$ corresponding to ζ'_{λ_1} . Similarly define $s_{2\lambda_2}$ in $W(\boldsymbol{a}_2, \boldsymbol{a}_{2\lambda_2})$ for $\lambda_2 \in I_2$. For simplicity, let s' denote $s_{1\alpha_0 1} \in W(\boldsymbol{a}_1, \boldsymbol{a})$, i.e., s' corresponds to $\zeta'_{\alpha_0 1}$. We have $s_{\lambda_1} = s_{1\lambda_1} s'^{-1}$. For simplicity, write P_{λ_1} and P_{λ_2} for $P_{1\lambda_1}$ and $P_{2\lambda_1}$ respectively $(\lambda_1 \in I_1 , \lambda_2 \in I_2)$.

In the first case, the only nonzero constant terms of $F^-(g)$ are the ones with respect to P_{λ_1} $(\lambda_1 \in I_1)$. It is easily checked that

$$W(\boldsymbol{a}, \boldsymbol{a}) = \{1, s's''\}$$

where $s'' = ss'^{-1} \in W(\boldsymbol{a}, \boldsymbol{a}_1)$, and

$$W(\boldsymbol{a}, \boldsymbol{a}_{\lambda_1}) = \begin{cases} \{s_{\lambda_1'}, s_{1\lambda_1'}s''\} & \text{for} \quad \lambda_1 = \lambda_1' = \alpha_0\beta \quad (1 \le \beta \le l_{\alpha_0}) \\ \\ \{s_{\lambda_1''}, s_{1\lambda_1''}s''\} & \text{for} \quad \lambda_1 = \lambda_1'' \in I_1 \setminus \{\alpha_0 1, \dots, \alpha_0 l_{\alpha_0}\} . \end{cases}$$

We have from the equation (6.5.2′) that

$$\text{(6.5.7)} \qquad F^{-\lambda_1'}(\Lambda, g) = \sum_n \phi^-(1\lambda_1', s_{1\lambda_1'}s'', n \mid \Lambda) e^{(-s_{1\lambda_1'}s''\Lambda - \rho_{\lambda_1'})(H_{\lambda_1'}(g))}$$

$$\times \sigma(k)\Psi_{1\lambda_1',\, s_{1\lambda_1'}s'',n}(m_{\lambda_1'}) ,$$

because $\phi^-(1\lambda_1', s_{\lambda_1'} , n) = 0$ by construction; and

$$\text{(6.5.8)} \qquad F^{-\lambda_1''}(\Lambda, g) = \sum_n \phi^-(1\lambda_1'', s_{1\lambda_1''}s'', n \mid \Lambda) e^{(-s_{1\lambda_1''}s''\Lambda - \rho_{\lambda_1''})(H_{\lambda_1''}(g))}$$

$$\times \sigma(k)\Psi_{1\lambda_1'',\, s_{1\lambda_1''}s'',\, n}(m_{\lambda_1''}) ,$$

because $\phi^{-}(1\lambda_1'', s_{\lambda_1''}, n) = 0$ by Proposition 6.4.1.

In the second case, the nonzero constant terms of $F^{-}(g)$ are the ones with respect to P_{λ_1}, P_{λ_2}. It is easily checked that

$$W(\boldsymbol{a}, \boldsymbol{a}_{\lambda_1}) = \{s_{\lambda_1}\},$$

and

$$W(\boldsymbol{a}, \boldsymbol{a}_{\lambda_2}) = \{s_{2\lambda_2}s''\}$$

where $s'' = ss'^{-1} \in W(\boldsymbol{a}, \boldsymbol{a}_2)$. We have from the equation (6.5.2′) that

$$F^{-\lambda_1'}(\Lambda, g) = 0\,, \tag{6.5.9}$$

because $\phi^{-}(1\lambda_1', s_{\lambda_1'}, n) = 0$ by construction;

$$F^{-\lambda_1''}(\Lambda, g) = 0\,, \tag{6.5.10}$$

because $\phi^{-}(1\lambda_1'', s_{\lambda_1''}, n) = 0$ by Proposition 6.5.1; and

$$F^{-\lambda_2}(\Lambda, g) = \sum_n \phi^{-}(2\lambda_2, 2_{2\lambda_2}s'', n \mid \Lambda)e^{(s_{2\lambda_2}s''\Lambda-\rho_{\lambda_2})(H_{\lambda_2}(g))} \tag{6.5.11}$$

$$\times\, \sigma(k)\Psi_{2\lambda_2,\, s_{2\lambda_2}s'',n}(m_{\lambda_2})\,.$$

Let $H \in \boldsymbol{a}$ be such that $\alpha(H) = 1$, where α is the simple root of $(\boldsymbol{p}, \boldsymbol{a})$. We now present the generalization of Lemma 4.3.6.

<u>Lemma 6.5.1.</u> Let i be either 1 or 2, according as whether we are in the first or the second case. If $\Lambda \in (\boldsymbol{a}_C^*)^+ \setminus \hat{\alpha}^{-1}(\text{spec } \mathbf{K}_\alpha)$ is such that $[(\text{Re } \Lambda)(H) - \rho(H)] + [\rho(H) - \rho_i(H_i)] > 0$, then $F^{-}(g)$ is bounded.

Proof. Let S_0, S be the Siegel domains, and F be the fundamental domain of Sections 6.3 and 6.4. We have $F \subset \Xi\, S_0 \subset \Xi\, S^o$, $\Xi = \{\zeta_v\}$. Since $F^-(g)$ is Γ-automorphic, slowly increasing, a σ-function and an eigenfunction for all of $\mathbf{Z}(g)$, $F^- - \sum_{rv}[\,\delta_{rv}F^{-rv}\,]_F$ is bounded on G. Write F^- as

$$(\,F^- - \sum_{rv}[\,\delta_{rv}F^{-rv}\,]_F\,) + \sum_{rv}[\,\delta_{rv}F^{-rv}\,]_F\,.$$

We demonstrate only for the first case, the second case can be handled analogously.

For $\sum_{rv}[\,\delta_{rv}F^{-rv}\,]_F$, we need only consider $\delta_{1\lambda_1}(g)F^{-\lambda_1}(\Lambda, g)$ for $g \in \Xi\, S^o$ $(\lambda_1 \in I_1)$. From equations (6.5.7) and (6.5.8), we have

$$\delta_{1\lambda_1}(g)F^{-\lambda_1}(\Lambda, g) = \sum_n \delta_{1\lambda_1}(g)\phi^-(1\lambda_1, s_{1\lambda_1}s'', n \mid \Lambda)$$

$$\times\, e^{(-s_{1\lambda_1}s''\Lambda - \rho_{\lambda_1})(H_{\lambda_1}(g))}\sigma(k)\Psi_{1\lambda_1,\, s_{1\lambda_1}s'',n}(m_{\lambda_1})\,.$$

Since $\delta_{1\lambda_1}(g) = 0$ for $g \notin \zeta_{\lambda_1}S^o$, we can replace $g \in \Xi\, S^o$ by $\zeta_{\lambda_1}g$ with $g \in S^o$. We need only show that the exponential

$$e^{(-s_{1\lambda_1}s''\Lambda - \rho_{\lambda_1})(H_{\lambda_1}(\zeta_{\lambda_1}g))}$$

is bounded for $g \in S^o$. Now

$$(-s_{1\lambda_1}s''\Lambda - \rho_{\lambda_1})(H_{\lambda_1}(\zeta_{\lambda_1}g)) = (-s''\Lambda - s_{1\lambda_1}^{-1}\rho_{\lambda_1})({}^{(\zeta_{\lambda_1}'^{-1})}H_{\lambda_1}(\zeta_{\lambda_1}g))$$

$$= (-s''\Lambda - s_{1\lambda_1}^{-1}\rho_{\lambda_1})(H_1(\zeta_{\lambda_1}'^{-1}\zeta_{\lambda_1}g))$$

$$= (-s''\Lambda - \rho_1)(H_1(\zeta g))\,,$$

where $\zeta = \zeta_{\lambda_1}'^{-1}\zeta_{\lambda_1} \in P_0$. Since $\zeta \in P_0 \subset P_1$, $H_1(\zeta g) = H_1(\zeta) + H_1(g)$, so

$$(-s''\Lambda - \rho_1)(H_1(\zeta g)) = (-s''\Lambda - \rho_1)(H_1(\zeta)) + (-s''\Lambda - \rho_1)(H_1(g))\,.$$

Write $H_1(g)$ as xH_1, where $H_1 \in \boldsymbol{a}$ and $\alpha_1(H_1) > 0$, $x \in \mathbf{R}$. We have $\alpha_1(H_1(g)) = x$. Refer to the proof of Lemma 4.3.6, and note that x is bounded from above for $g = S^o$. Moreover,

$$\begin{aligned}(-s''\Lambda - \rho_1)(H_1(g)) &= (-s''\Lambda - \rho_1)(xH_1) \\ &= -x(s''\Lambda)(H_1) - x\rho_1(H_1) \\ &= -x(ss'^{-1}\Lambda)(H_1) - x\rho_1(H_1) \\ &= x\Lambda(H) - x\rho_1(H_1),\end{aligned}$$

since $H = s'H_1$. Thus

$$(-s''\Lambda - \rho_1)(H_1(g)) = x\{\, [\Lambda(h) - \rho(H)] + [\rho(H) - \rho_1(H_1)]\, \}.$$

Therefore,

$$\left| e^{(-s_{1\lambda_1}s''\Lambda - \rho_{\lambda_1})(H_{\lambda_1}(\zeta_{\lambda_1}g))} \right| = e^{x\{\, [(\mathrm{Re}\Lambda)(H) - \rho(H)] + [\rho(H) - \rho_1(H_1)]\, \}},$$

which is bounded for $g \in S^o$, provided

$$[(\mathrm{Re}\Lambda)(H) - \rho(H)] + [\rho(H) - \rho_1(H_1)] > 0. \quad \square$$

The rest of the material of Chapter 4 requires only little or no modifications to become valid in general.

6.6. Functional Equations.

Before we state and prove the functional equations, we would do well to review the combinatorics of maximal cuspidal subgroups. Besides, we need to develop the formalism of the last four sections further.

In this section we concern ourselves exclusively with associate cuspidal subgroups. We are therefore looking at the last maximal cuspidal subgroups P_{jv_j} for

$$\nu_j \in I_j = \{11, \ldots, 1l_1; 21, \ldots, 2l_2; \cdots; d(r)1, \ldots, d(r)l_{d(r)}\},$$

where j = 1 or j = 1, 2, according as we are in the first or second case of Section 6.5.

Recall that

$$P_{j\nu_j} = {}^{\zeta_{\nu_j}}(P_j), \qquad \Xi = \{\zeta_{\nu_j}\}, \qquad G = \Gamma\,\Xi\,S_0.$$

The subset $\{\zeta_{\nu_j} : \nu_j \in \underline{I}_j\}$ of Ξ, where $\underline{I}_j = \{11, 21, \ldots, d(j)1\}$, is a complete set of representatives for $\Gamma \backslash \Xi / (P_j)_Q$, and $P_{j\alpha\beta} = {}^{\gamma_{\alpha\beta}}(P_{j\alpha 1})$ $(1 \le \alpha < d(j),\ 1 < \beta \le l_\alpha)$ for some $\gamma_{\alpha\beta} \in \Gamma$. For $\zeta_{\nu_j} = \zeta'_{\nu_j}\zeta''_{\nu_j}$ where $\zeta'_{\nu_j} \in K$, and $\zeta''_{\nu_j} \in P_0 \subset P_j$, $P_{j\nu_j} = N_{j\nu_j}A_{j\nu_j}M_{j\nu_j}$ with

$$(P_{j\nu_j}; N_{j\nu_j}, A_{j\nu_j}, M_{j\nu_j}) = {}^{\zeta'_{\nu_j}}(P_j; N_j, A_j, M_j).$$

The Weyl group element $s_{j\nu_j}$ in $W(\mathfrak{a}_j, \mathfrak{a}_{j\nu_j})$ corresponds to ζ'_{ν_j}.

In the first case, $P_1 = P_{111}$ is a maximal standard cuspidal subgroup which has only itself as associate among all the maximal standard cuspidal subgroups. For simplicity, drop all subscripts of j = 1. Thus $W(\mathfrak{a}, \mathfrak{a}) = \{1, s\}$. Let ${}_{\nu'}s_\nu \in W'(\mathfrak{a}_\nu, \mathfrak{a}_{\nu'})$ be $s_{\nu'} s s_\nu^{-1}$ for $\nu, \nu' \in I$. Then

$$W(\mathfrak{a}_\nu, \mathfrak{a}_{\nu'}) = \{s_{\nu'} s_\nu^{-1}, {}_{\nu'}s_\nu\}.$$

In the second case, $P_1 = P_{111}$ and $P_2 = P_{211}$ are two associate maximal standard cuspidal subgroups. We have $W(\mathfrak{a}_1, \mathfrak{a}_2) = \{s\}$. Let ${}_{\nu_2}s_{\nu_1} \in W(\mathfrak{a}_{\nu_1}, \mathfrak{a}_{\nu_2})$ be $s_{\nu_2} s s_{\nu_1}^{-1}$, and ${}_{\nu_1}s_{\nu_2} \in W(\mathfrak{a}_{\nu_2}, \mathfrak{a}_{\nu_1})$ be $s_{\nu_1} s^{-1} s_{\nu_2}^{-1}$ for $\nu_1 \in I_1$, $\nu_2 \in I_2$ (of course, s_{ν_1}, s_{ν_2} denote $s_{1\nu_1}$, $s_{2\nu_2}$). Then

$$W(\mathfrak{a}_{\nu_1}, \mathfrak{a}_{\nu_2}) = \{{}_{\nu_2}s_{\nu_1}\}, \qquad W(\mathfrak{a}_{\nu_2}, \mathfrak{a}_{\nu_1}) = \{{}_{\nu_1}s_{\nu_2}\}.$$

Fix a character χ of $\mathbf{Z}(M_1)$ for what follows.

In the first case, let

$$L_\nu = {}^0L^2(\Gamma_\nu \setminus M_\nu, \sigma_\nu, {}^{s_\nu}\chi) + {}^0L^2(\Gamma_\nu \setminus M_\nu, \sigma_\nu, {}^{\nu s_{11}}\chi) \qquad (\nu \in \underline{I}),$$

and

$$L = \prod_{\nu \in \underline{I}} L_\nu .$$

Denote by $\underline{\boldsymbol{a}_C^*}$ the subset of $\prod_{\nu \in \underline{I}} (\boldsymbol{a}_\nu)_C^*$:

$$\{ (\Lambda_\nu) : \Lambda_\nu = s_\nu \Lambda \quad (\Lambda \in \boldsymbol{a}_C^*) \quad \forall\, \nu \in \underline{I} \} .$$

For $\underline{\Phi} = (\Phi_\nu) \in L$, $\underline{\Lambda} = (\Lambda_\nu) \in \underline{\boldsymbol{a}_C^*}$, define the *complete Eisenstein series*

$$E(\underline{\Lambda}, \underline{\Phi}, g)$$

to be

$$\sum_{\nu \in \underline{I}} E_\nu(\Lambda_\nu, \Phi_\nu, g) \qquad (g \in G) .$$

Each summand is a sum of two usual Eisenstein series (cf. Section 6.2), and is thus meromorphic on the entire $(\boldsymbol{a}_\nu)_C^*$. Because for $\underline{\Lambda} \in \underline{\boldsymbol{a}_C^*}$ the dual spaces $(\boldsymbol{a}_\nu)_C^*$ are all parametrized by $\Lambda \in \boldsymbol{a}_C^*$, the complete Eisenstein series can be regarded as meromorphic on the entire $\boldsymbol{a}_C^*$. In all that follows it will always be assumed that $\Lambda \in \boldsymbol{a}_C^*$ avoids all poles of the functions involved.

The *complete intertwining operator* is a matrix whose entries are intertwining operators of the previous chapters:

$$M(\underline{\Lambda}) = [M({}_\nu s_{\nu'}, \Lambda_{\nu'})]_{\nu,\nu' \in \underline{I}} \qquad (\underline{\Lambda} \in \underline{\boldsymbol{a}_C^*}) .$$

Strictly speaking, we have dealt only with intertwining operators of the kind from ${}^0L^2(\Gamma' \setminus M', \sigma', \chi)$ into ${}^0L^2(\Gamma'' \setminus M'', \sigma'', {}^s\chi)$ for a character χ of $\mathbf{Z}(M')$, M' and M'' being Levi components of two associate maximal cuspidal subgroups. However, one can let a finite set ξ of characters of $\mathbf{Z}(M)$ take the place of χ, and define in the obvious manner an intertwining operator from the direct sum $\sum_{\chi \in \xi} {}^0L^2(\Gamma' \setminus M', \sigma', \chi)$ into the direct sum $\sum_{\chi \in \xi} {}^0L^2(\Gamma'' \setminus M'', \sigma'', {}^s\chi)$. Langlands' formula (Theorem 6.4.1) for constant terms remains valid. Through matrix multiplication $M(\underline{\Lambda})$ acts as a linear transformation on L:

$$M(\underline{\Lambda}) : L \to L : \underline{\Phi} \mapsto M(\underline{\Lambda})(\Phi_\nu),$$

where we regard (Φ_ν) as a column matrix.

In the second case, let

$$L_{\nu_1} = {}^0L^2(\Gamma_{\nu_1} \setminus M_{\nu_1}, \sigma_{\nu_1}, {}^{s_{\nu_1}}\chi) \qquad (\nu_1 \in \underline{I_1}),$$

$$L_{\nu_2} = {}^0L^2(\Gamma_{\nu_2} \setminus M_{\nu_2}, \sigma_{\nu_2}, {}^{s_{\nu_2}}\chi) \qquad (\nu_2 \in \underline{I_2}),$$

and

$$L = \prod_{\nu_1 \in \underline{I_1}} L_{\nu_1} \times \prod_{\nu_2 \in \underline{I_2}} L_{\nu_2}.$$

Denote by $\underline{\mathfrak{a}_C^*}$ the subset of

$$\prod_{\nu_1 \in \underline{I_1}} (\mathfrak{a}_{\nu_1})_C^* \times \prod_{\nu_2 \in \underline{I_2}} (\mathfrak{a}_{\nu_2})_C^* :$$

$$\{ (\Lambda_{\nu_1}, \Lambda_{\nu_2}) : \Lambda_{\nu_1} = s_{\nu_1}\Lambda_1 , \Lambda_{\nu_2} = s_{\nu_2}\Lambda_2 \quad (\Lambda_1 \in (\mathfrak{a}_1)_C^* , \Lambda_2 = -s\Lambda_1)$$

$$\forall \, \nu_1 \in \underline{I_1} , \nu_2 \in \underline{I_2} \}.$$

For $\underline{\Phi} = (\Phi_{\nu_1}, \Phi_{\nu_2}) \in L$, $\underline{\Lambda} = (\Lambda_{\nu_1}, \Lambda_{\nu_2}) \in \underline{\mathfrak{a}_C^*}$, define the complete Eisenstein series

$$E(\underline{\Lambda}, \phi, g)$$

to be

$$\sum_{\nu_1 \in \underline{I_1}} E_{\nu_1}(\Lambda_{\nu_1}, \Phi_{\nu_1}, g) + \sum_{\nu_2 \in \underline{I_2}} E_{\nu_2}(\Lambda_{\nu_2}, \Phi_{\nu_2}, g) \qquad (g \in G).$$

The complete intertwining operator is a matrix whose entries are intertwining operators of the previous chapters, and is given as the following block matrix:

$$\underline{M}(\underline{\Lambda}) = \begin{bmatrix} & [M(v_1 s_{v_2} \mid \Lambda_{v_2})]_{v_1 \in \underline{I_1},\, v_2 \in \underline{I_2}} \\ [M(v_2 s_{v_1} \mid \Lambda_{v_1})]_{v_2 \in \underline{I_2},\, v_1 \in \underline{I_1}} & \end{bmatrix} .$$

It acts as a linear transformation on $\underline{L}$.

We can now state the functional equations.

Theorem 6.6.1. In either case, we have the following functional equations of the Eisenstein series and the intertwining operator:

$$\text{(6.6.1)} \qquad \underline{E}(-\underline{\Lambda}, \underline{\Phi}) = \underline{E}(\underline{\Lambda}, \underline{M}(-\underline{\Lambda})\underline{\Phi}) \qquad (\underline{\Lambda} \in \underline{a_C^*}) \qquad \text{for all } \underline{\Phi} \in \underline{L},$$

$$\text{(6.6.2)} \qquad \underline{M}(\underline{\Lambda})\underline{M}(-\underline{\Lambda}) = I \qquad (\underline{\Lambda} \in \underline{a_C^*}) .$$

Before giving the proof let us first see that equations (6.6.1) and (6.6.2) respectively reduce to equations (5.2.3) and (5.2.4) of Chapter 5 when there is only one cusp. This is obvious in the first case. In the second case, the equation (6.6.1) becomes

$$E_1(-\Lambda_1, \Phi_1, g) + E_2(-\Lambda_2, \Phi_2, g) = E_2(\Lambda_2, M(s \mid -\Lambda_1)\Phi_1, g) + E_1(\Lambda_1, M(s^{-1} \mid -\Lambda_2)\Phi_2, g) ,$$

and equation (6.6.2) becomes

$$\begin{bmatrix} & M(s^{-1} \mid \Lambda_2) \\ M(s \mid \Lambda_1) & \end{bmatrix} \begin{bmatrix} & M(s^{-1} \mid -\Lambda_2) \\ M(s \mid -\Lambda_1) & \end{bmatrix} = \begin{bmatrix} I & \\ & I \end{bmatrix} ,$$

which is equivalent to

$$\begin{cases} M(s^{-1} \mid \Lambda_2) M(s \mid -\Lambda_1) = I \\ M(s \mid \Lambda_1) M(s^{-1} \mid -\Lambda_2) = I . \end{cases}$$

Note that $\Phi_1 \in {}^0L^2(\Gamma_{M_1} \backslash M_1, \sigma_{M_1}, \chi)$, $\Phi_2 \in {}^0L^2(\Gamma_{M_2} \backslash M_2, \sigma_{M_2}, {}^s\chi)$, $\Lambda_2 = -s\Lambda_1$. These clearly imply equations (5.2.3) and (5.2.4). (They are apparently stronger only because the formulation here is completely symmetric with respect to P_1 and P_2.)

The derivation of equations (6.6.1) and (6.6.2) will follow the uniquness development of the last section, similar to what was done in Section 5.2. However, this time the combinatorics is somewhat involved. We need a technical lemma, which has in effect been demonstrated in the course of proving Lemma 6.5.1. We state it here as Lemma 6.6.1.

Lemma 6.6.1. In the first case, for $\Lambda \in \mathfrak{a}_{\mathbb{C}}^*$ such that $(\mathrm{Re}\,\Lambda)(H) - \rho(H) > 0$,

$$e^{(-s_\lambda s\Lambda - \rho_\lambda)(H_\lambda(\zeta_\lambda g))} \qquad (\lambda \in I)$$

are bounded for $g \in S^o$.

In the second case, if in addition $\Lambda_1 \in (\mathfrak{a}_1)_{\mathbb{C}}^*$ is such that $[(\mathrm{Re}\,\Lambda_1)(H_1) - \rho_1(H_1)] + [\rho_1(H_1) - \rho_2(H_2)] > 0$, then

$$e^{(s_{\lambda_1}\Lambda_1 - \rho_{\lambda_1})(H_{\lambda_1}(\zeta_{\lambda_1} g))} \qquad (\lambda_1 \in I_1)$$

and

$$e^{(-s_{\lambda_2} s\Lambda_1 - \rho_{\lambda_2})(H_{\lambda_2}(\zeta_{\lambda_2} g))} \qquad (\lambda_2 \in I_2)$$

are bounded for $g \in S^o$.

Proof of Theorem 6.6.1. We prove the equation (6.6.1) first. Begin with the first case. Referring to the definition of $E(\underline{\Lambda}, \underline{\Phi}, g)$, the functional equation (6.6.1) is equivalent to

$$\sum_{\nu \in \underline{I}} E_\nu(-\Lambda_\nu, \Phi_\nu, g) = \sum_{\nu \in \underline{I}} \sum_{\nu' \in \underline{I}} E_{\nu'}(\Lambda_{\nu'}, M({}_{\nu'}s_\nu \,|\, -\Lambda_\nu)\Phi_\nu, g).$$

So for the equation (6.6.1), it suffices to prove that

(6.6.3) $$E_\nu(-\Lambda_\nu, \Phi_\nu, g) = \sum_{\nu' \in \underline{I}} E_{\nu'}(\Lambda_{\nu'}, M(_{\nu'}s_\nu \mid -\Lambda_\nu)\Phi_\nu, g) \qquad (\nu \in \underline{I}).$$

Recall that $\Lambda_\nu = s_\nu\Lambda$ $(\Lambda \in \boldsymbol{a}_C^*)$ for all $\nu \in \underline{I}$. Let

$$E(\Lambda, g) = E_\nu(-\Lambda_\nu, \Phi_\nu, g) - \sum_{\nu' \in \underline{I}} E_{\nu'}(\Lambda_{\nu'}, M(_{\nu'}s_\nu \mid -\Lambda_\nu)\Phi_\nu, g).$$

For a given $\nu \in \underline{I}$, this is Γ-automorphic, slowly increasing, a σ-function, and $Z(G)$-finite. Moreover, it is an eigenfunction of the Casimer operator ω with the eigenvalue a nonconstant entire function of Γ, because $E_\nu(-\Lambda_\nu, \Phi_\nu, g)$ and each of $E_{\nu'}(\Lambda_{\nu'}, M(_{\nu'}s_\nu \mid -\Lambda_\nu)\Phi_\nu, g)$ share this same eigenvalue (cf. Section 5.2; Harish-Chandra [2, p. 88]). Write E as

$$\Big(E - \sum_{r\lambda} [\, \delta_{r\lambda} E^{r\lambda} \,]_F\Big) + \sum_{r\lambda} [\, \delta_{r\lambda} E^{r\lambda} \,]_F .$$

For $\Lambda \in \boldsymbol{a}_C^*$, $E - \sum_{r\lambda} [\, \delta_{r\lambda} E^{r\lambda} \,]_F$ is bounded on G. As in Section 5.2, we shall prove that there exists an open subset of $\boldsymbol{a}_C^*$ such that for every Λ in the set $\sum_{r\lambda}[\, \delta_{r\lambda} E^{r\lambda} \,]_F$ is a bounded function on G. There is then an open subset of $\boldsymbol{a}_C^*$ such that for every Λ in the set the function E is a bounded eigenfunction of the Casimir operator with nonreal eigenvalue. By Theorem 4.3.1, for all such Λ, $E(g) = 0$ $(g \in G)$. The equation (6.6.3) follows from this because both sides are known to be entire meromorphic on $\boldsymbol{a}_C^*$.

Of course, $\sum_{r\lambda} [\, \delta_{r\lambda} E^{r\lambda} \,]_F$ is just $\sum_{\lambda \in I} [\, \delta_\lambda E^\lambda \,]_F$ (we have written δ_λ for $\delta_{1\lambda}$). Note that it suffices to bound $\delta_\lambda(\zeta_\lambda g) E^\lambda(\zeta_\lambda g)$ for $g \in S^o$. Recall that $W(\boldsymbol{a}_\nu, \boldsymbol{a}_\lambda) = \{s_\lambda s_\nu^{-1}, {}_\lambda s_\nu\}$. Suppose at first $\lambda \in \underline{I}$. Regarding the left-hand side of the equation (6.6.3), we have from Langlands' formula that for $\zeta_\lambda g = n_\lambda a_\lambda m_\lambda k$, where $n_\lambda \in N_\lambda$, $a_\lambda \in A_\lambda$, $m_\lambda \in M_\lambda$, $k \in K$,

$$E_\nu^\lambda(-\Lambda_\nu, \Phi_\nu, \zeta_\lambda g) = e^{(s_\lambda s_\nu^{-1}\Lambda_\nu-\rho_\lambda)(H_\lambda(\zeta_\lambda g))}\sigma(k)(M(s_\lambda s_\nu^{-1}\,|-\Lambda_\nu)\Phi_\nu)(m_\lambda)$$
$$+\, e^{({}_\lambda s_\nu\Lambda_\nu-\rho_\lambda)(H_\lambda(\zeta_\lambda(g))}\,\sigma(k)(M({}_\lambda s_\nu\,|-\Lambda_\nu)\Phi_\nu)(m_\lambda)\,.$$

The operator $M(s_\lambda s_\nu^{-1}\,|-\Lambda_\nu)$ equals to the identity transformation if $\lambda = \nu$, and by Proposition 6.5.2, it is zero if $\lambda \neq \nu$. Thus

$$(6.6.4)\quad E_\nu^\lambda(-\Lambda_\nu, \Phi_\nu, \zeta_\lambda g) = \begin{cases} e^{(-s_\nu s\Lambda-\rho_\nu)(H_\nu(\zeta_\nu g))}\sigma(k)\Phi_\nu(m_\nu) \\ \qquad +\, \underline{e^{(-\Lambda_\nu-\rho_\nu)(H_\nu(\zeta_\nu g))}\sigma(k)(M({}_\nu s_\nu\,|-\Lambda_\nu)\Phi_\nu)(m_\nu)} \\ \qquad\qquad \text{if } \lambda = \nu \\ \underline{e^{(-\Lambda_\lambda-\rho_\lambda)(H_\lambda(\zeta_\lambda))}\sigma(k)(M({}_\lambda s_\nu\,|-\Lambda_\nu)\Phi_\nu)(m_\lambda)} \\ \qquad\qquad \text{if } \lambda \neq \nu\,. \end{cases}$$

Restrict $\Lambda \in \boldsymbol{a}_{\mathbb{C}}^*$ such that $(\mathrm{Re}\,\Lambda)(H) - \rho(H) > 0$. By Lemma 6.6.1, the term with the exponential $e^{(-s_\nu s\Lambda-\rho_\nu)(H_\nu(\zeta_\nu g))}$ is bounded for $g \in S^0$. On the other hand, for the right-hand side of the equation (6.6.3), the same reasoning shows that

$$(6.6.5)\quad E_{\nu'}^\lambda(\Lambda_{\nu'}, M({}_{\nu'}s_\nu\,|-\Lambda_\nu)\Phi_\nu, \zeta_\lambda g) = \begin{cases} \underline{e^{(-\Lambda_{\nu'}-\rho_{\nu'})(H_{\nu'}(\zeta_{\nu'}(g))}\sigma(k)(M({}_{\nu'}s_\nu\,|-\Lambda_\nu)\Phi_\nu)(m_{\nu'})} \\ \quad +\, e^{(-s_{\nu'}s\Lambda-\rho_{\nu'})(H_{\nu'}(\zeta_{\nu'}g))} \\ \qquad \times\, \sigma(k)(M({}_{\nu'}s_{\nu'}\,|\,\Lambda_{\nu'})M({}_{\nu'}s_\nu\,|-\Lambda_\nu)\Phi_\nu)(m_{\nu'}) \\ \qquad\qquad \text{if } \lambda = \nu' \\ e^{(-s_\lambda s\Lambda-\rho_\lambda)(H_\lambda(\zeta_\lambda g))} \\ \qquad \times\, \sigma(k)(M({}_\lambda s_{\nu'}\,|\,\Lambda_{\nu'})M({}_{\nu'}s_\nu\,|-\Lambda_\nu)\Phi_\nu(m_\lambda) \\ \qquad\qquad \text{if } \lambda \neq \nu'\,. \end{cases}$$

By Lemma 6.6.1, the terms with the exponentials $e^{(-s_{\nu'}s\Lambda-\rho_{\nu'})(H_{\nu'}(\zeta_{\nu'}g))}$ and $e^{(-s_\lambda s\Lambda-\rho_\lambda)(H_\lambda(\zeta_\lambda g))}$ are bounded. In the above, we have underlined those terms which may not be bounded. Comparing the two sets of formulae, we see that for any $\lambda \in \underline{I}$,

$$\delta_\lambda[\, E_\nu^\lambda(-\Lambda_\nu, \Phi_\nu, \zeta_\lambda g) - E_\lambda^\lambda(\Lambda_\lambda, M({}_\lambda s_\nu\,|-\Lambda_\nu)\Phi_\nu, \zeta_\nu g)\,]$$

is bounded, as well as the remaining part of

$$\delta_\lambda \sum_{\nu' \in \underline{I} \setminus \{\lambda\}} E_{\nu'}^{\lambda}(\Lambda_{\nu'}, M({}_{\nu'}s_\nu \,|\, -\Lambda_\nu)\Phi_\nu, \zeta_\lambda g) .$$

Now more generally, suppose $\lambda_0 = \alpha_0 1 \in \underline{I}$ and $\lambda = \alpha_0\beta \in I$ $(1 \le \beta \le l_{\alpha_0})$. Since ${}^{\gamma_\lambda}(P_{\lambda_0}) = P_\lambda$ for some $\gamma_\lambda \in \Gamma$, Proposition 6.5.1 asserts that

$$\delta_\lambda E_\nu^{\lambda}(-\Lambda_\nu, \Phi_\nu, \zeta_\lambda g) = \delta_\lambda E_\nu^{\lambda_0}(-\Lambda_\nu, \Phi_\nu, \gamma_\lambda^{-1}\zeta_\lambda g) .$$

So the equation (6.6.4) can be used to express $E_\nu^{\lambda}(-\Lambda_\nu, \Phi_\nu, \zeta_\lambda g)$. For this matter note that

$${}^{\zeta_{\lambda_0}}(P_1) = P_{\lambda_0} \quad \text{and} \quad {}^{\zeta_\lambda}(P_1) = P_\lambda$$

implies

$${}^{\zeta_\lambda \zeta_{\lambda_0}^{-1}}(P_{\lambda_0}) = P_\lambda = {}^{\gamma_\lambda}(P_{\lambda_0}) ,$$

from which we have $\gamma_\lambda^{-1}\zeta_\lambda = p\zeta_{\lambda_0}$ for $p \in P_{\lambda_0}$, so that

$$H_{\lambda_0}(\gamma_\lambda^{-1}\zeta_\lambda g) = H_{\lambda_0}(p\zeta_{\lambda_0} g) = H_{\lambda_0}(p) + H_{\lambda_0}(\zeta_{\lambda_0} g) .$$

Thus the growth of the various terms is exactly as discussed after the equation (6.6.4). Like before,

$$\delta_\lambda[E_\nu^{\lambda}(-\Lambda_\nu, \Phi_\nu, \zeta_\lambda g) - E_{\lambda_0}^{\lambda}(\Lambda_{\lambda_0}, M({}_{\lambda_0}s_\nu \,|\, -\Lambda_\nu)\Phi_\nu, \zeta_\lambda g)]$$

is bounded, as well as the remaining part of

$$\delta_\lambda \sum_{\nu' \in \underline{I} \setminus \{\lambda_0\}} E_{\nu'}^{\lambda}(\Lambda_{\nu'}, M({}_{\nu'}s_\nu \,|\, -\Lambda_\nu)\Phi_\nu, \zeta_\lambda g) .$$

We have therefore proved that, for $\Lambda \in \boldsymbol{a}_{\mathbb{C}}^*$ such that $(\mathrm{Re}\,\Lambda)(H) - \rho(H) > 0$, $\sum_{\lambda \in I} [\, \delta_\lambda E^\lambda \,]_F$ is bounded. As discussed at the beginning of our proof, this implies the equation (6.6.3).

To prove the equation (6.6.1) in the second case, one follows the same line of reasoning except for the slightly different combinatorial set-up. Referring to the corresponding definition of $E(\underline{\Lambda}, \underline{\Phi}, g)$, in the second case the functional equation (6.6.1) is equivalent to

$$\sum_{v_1 \in \underline{I_1}} E_{v_1}(-\Lambda_{v_1}, \Phi_{v_1}, g) + \sum_{v_2 \in \underline{I_2}} E_{v_2}(-\Lambda_{v_2}, \Phi_{v_2}, g)$$

$$= \sum_{v_1 \in \underline{I_1}} \sum_{v_2 \in \underline{I_2}} E_{v_2}(\Lambda_{v_2}, M({}_{v_2}s_{v_1} \,|\, \Lambda_{v_1})\Phi_{v_1}, g)$$

$$+ \sum_{v_2 \in \underline{I_2}} \sum_{v_1 \in \underline{I_1}} E_{v_1}(\Lambda_{v_1}, M({}_{v_1}s_{v_|}\Lambda_{v_2})\Phi_{v_2}, g) .$$

Because of symmetry, it suffices to demonstrate that

$$\text{(6.6.6)} \qquad E_{v_1}(-\Lambda_{v_1}, \Phi_{v_1}, g) = \sum_{v_2 \in \underline{I_2}} E_{v_2}(\Lambda_{v_2}, M({}_{v_2}s_{v_1} \,|\, \Lambda_{v_1})\Phi_{v_1}, g) \text{ for } v_1 \in \underline{I_1} .$$

Recall that for all $v_1 \in \underline{I_1}$, $v_2 \in \underline{I_2}$, $\Lambda_{v_1} = s_{v_1}\Lambda_1$, $\Lambda_{v_2} = s_{v_2}\Lambda_2$, where $\Lambda_1 \in (\boldsymbol{a}_1)^*_C$, $\Lambda_2 = -\,s\Lambda_1$.

Let

$$E(\Lambda_1, g) = E_{v_1}(-\Lambda_{v_1}, \Phi_{v_1}, g) - \sum_{v_2 \in \underline{I_2}} E_{v_2}(\Lambda_{v_2}, M({}_{v_2}s_{v_1} \,|\, \Lambda_{v_1})\Phi_{v_1}, g) .$$

Now $\sum_{r\lambda} [\, \delta_{r\lambda} E^{r\lambda} \,]_F$ is $\sum_{\lambda_1 \in I_1} [\, \delta_{\lambda_1} E^{\lambda_1} \,]_F + \sum_{\lambda_2 \in I_2} [\delta_{\lambda_2} E^{\lambda_2}]_F$. Recall that $W(\boldsymbol{a}_{v_1}, \boldsymbol{a}_{\lambda_1}) = \{s_{\lambda_1} s_{v_1}^{-1}\}$, $W(\boldsymbol{a}_{v_1}, \boldsymbol{a}_{\lambda_2}) = \{{}_{\lambda_2}s_{v_1}\}$. Suppose at first $\lambda_1 \in \underline{I_1}$, $\lambda_2 \in \underline{I_2}$. Regarding the left-hand side of the equation (6.6.6), we have from Langlands' formula that

$$E_{v_1}^{\lambda_1}(-\Lambda_{v_1}, \Phi_{v_1}, \zeta_{\lambda_1} g) = e^{(s_{\lambda_1} s_{v_1}^{-1} \Lambda_{v_1} - \rho_{\lambda_1})(H_{\lambda_1}(\zeta_{\lambda_1} g))}$$

$$\times \sigma(k)(M(s_{\lambda_1} s_{v_1}^{-1} \,|\, -\Lambda_{v_1})\Phi_{v_1})(m_{\lambda_1})$$

$$= \begin{cases} e^{(s_{\nu_1}\Lambda_1-\rho_{\nu_1})(H_{\nu_1}(\zeta_{\nu_1}g))}\sigma(k)\Phi_{\nu_1}(m_{\nu_1}) & \text{if } \lambda_1=\nu_1 \\ 0 & \text{if } \lambda_1\neq\nu_1 , \end{cases}$$

and

$$E_{\nu_1}^{\lambda_2}(-\Lambda_{\nu_1}, \Phi_{\nu_1}, \zeta_{\lambda_2}g) = e^{(-_{\lambda_2}s_{\nu_1}\Lambda_{\nu_1}-\rho_{\lambda_2})(H_{\lambda_2}(\zeta_{\lambda_2}g))}$$

$$\times\,\sigma(k)(M(_{\lambda_2}s_{\nu_1} \mid -\Lambda_{\nu_1})\Phi_{\nu_1})(m_{\lambda_2})$$

$$= \underline{e^{(-\Lambda_{\lambda_2}-\rho_{\lambda_2})(H_{\lambda_2}(\zeta_{\lambda_2}g))}}$$

$$\underline{\times\,\sigma(k)(M(_{\lambda_2}s_{\nu_1} \mid -\Lambda_{\nu_1})\Phi_{\nu_1})(m_{\lambda_2})} .$$

Restrict $\Lambda_1 \in (\boldsymbol{a}_1)_C^*$ so that $(\mathrm{Re}\,\Lambda_1)(H_1) - \rho_1(H_1) > 0$. By Lemma 6.6.1, the term with the exponentials $e^{(s_{\nu_1}\Lambda_1-\rho_{\nu_1})(H_{\nu_1}(\zeta_{\nu_1}g))}$ is bounded for $g \in S^o$. Thus $E_{\nu_1}^{\lambda_1}(-\Lambda_{\nu_1}, \Phi_{\nu_1}, \zeta_{\lambda_1}g)$ is bounded for all $\lambda_1 \in \underline{I_1}$. On the other hand, for the right-hand side of the equation (6.6.6),

$$E_{\nu_2}^{\lambda_1}(\Lambda_{\nu_2}, M(_{\nu_2}s_{\nu_1} \mid \Lambda_{\nu_1})\Phi_{\nu_1}, \zeta_{\lambda_1}g) = e^{(s_{\lambda_1}\Lambda_1-\rho_{\lambda_1})(H_{\lambda_1}(\zeta_{\lambda_1}g))}$$

$$\times\,\sigma(k)(M_{\lambda_1}s_{\nu_1} \mid \Lambda_{\nu_2})M(_{\nu_2}s_{\nu_1} \mid \Lambda_{\nu_1})\Phi_{\nu_1})(m_{\nu_1}) ,$$

and

$$E_{\nu_2}^{\lambda_2}(\Lambda_{\nu_2}, M(_{\nu_2}s_{\nu_1} \mid \Lambda_{\nu_1})\Phi_{\nu_1}, \zeta_{\lambda_2}g)$$

$$= \begin{cases} \underline{e^{(-\Lambda_{\nu_2}-\rho_{\nu_2})(H_{\nu_2}(\zeta_{\nu_2}g))}\sigma(k)(M(_{\nu_2}s_{\nu_1} \mid \Lambda_{\nu_1})\Phi_{\nu_1})(m_{\nu_2})} & \text{if } \lambda_2=\nu_2 \\ 0 & \text{if } \lambda_2\neq\nu_2 . \end{cases}$$

Obviously, $[\,\delta_{\lambda_1}E^{\lambda_1}\,]_F$ is bounded for all $\lambda_1 \in \underline{I_1}$. Besides, for any $\lambda_2 \in \underline{I_2}$,

$$\delta_{\lambda_2}[\,E_{\nu_1}^{\lambda_2}(-\Lambda_{\nu_1}, \Phi_{\nu}, \zeta_{\lambda_2}g) - E_{\nu_2}^{\lambda_2}(\Lambda_{\lambda_2}, M(_{\lambda_2}s_{\nu_1} \mid \Lambda_{\nu_1})\Phi_{\nu_1}, \zeta_{\lambda_2}g)\,]$$

is bounded, as well as the remaining part of

$$\delta_{\lambda_2} \sum_{\nu_2 \in \underline{I_2} \setminus \{\lambda_2\}} E_{\nu_2}^{\lambda_2}(\Lambda_{\nu_2}, M(_{\nu_2}s_{\nu_1} \mid \Lambda_{\nu_1})\Phi_{\nu_1}, \zeta_{\lambda_2}g) .$$

The rest of the development for the equation (6.6.1) in the second case is clear from that in the first case.

In closing we deduce the other functional equation (6.6.2). It follows immediately from the functional equation (6.6.1) that

$$E(\underline{\Lambda}, \underline{\Phi}) = E(-\underline{\Lambda}, M(\underline{\Lambda})\underline{\Phi}) \qquad (\underline{\Lambda} \in \underline{a_C^*}) \qquad \text{for all } \underline{\Phi} \in L .$$

Together with equation (6.6.1), this implies that

$$(6.6.7) \qquad E(\underline{\Lambda}, \underline{\Phi}) = E(\underline{\Lambda}, M(-\underline{\Lambda})M(\underline{\Lambda})\underline{\Phi}) \qquad (\underline{\Lambda} \in \underline{a_C^*}) \qquad \text{for all } \underline{\Phi} \in L .$$

The function equation (6.6.2) follows from Equation (6.6.7) on account of the principle that

$$(6.6.8) \qquad E(\underline{\Lambda}, \underline{\Phi}^{(1)}) = E(\underline{\Lambda}, \underline{\Phi}^{(2)}) \qquad (\underline{\Lambda} \in \underline{a_C^*})$$

implies

$$\underline{\Phi}^{(1)} = \underline{\Phi}^{(2)} .$$

In the first case, say, the equation (6.6.8) is equivalent to

$$\sum_{\nu \in \underline{I}} E_\nu(\Lambda_\nu, \Phi_\nu^{(1)}, g) = \sum_{\nu \in \underline{I}} E_\nu(\Lambda_\nu, \Phi_\nu^{(2)}, g) .$$

To see why the principle is valid, for each $\nu \in \underline{I}$, consider the constant term with respect to P_ν of the two sides of the last equation. For $g \in N_0A_0(t)K$, and $\Lambda \in a_C^*$ such that $(\mathrm{Re}\,\Lambda)(H) - \rho(H) > 0$, they respectively have a unique growing term of

$$e^{(s_\nu s\Lambda - \rho_\nu)(H_\nu(\zeta_\nu g))}\sigma(k)\Phi_\nu^{(1)}(m_\nu) ,$$

and

$$e^{(s_\nu s\Lambda - \rho_\nu)(H_\nu(\zeta_\nu g))}\sigma(k)\Phi_\nu^{(2)}(m_\nu) .$$

One can equate these to deduce that $\Phi_\nu^{(1)} = \Phi_\nu^{(2)}$. This completes the proof of Theorem 6.6.1. □

REFERENCES

Arthur, J.

[1] Eisenstein series and the trace formula, Proc. Symp. Pure Math., A.M.S., Vol. XXXIII (1979), pp. 253-262.

Baily, W.

[1] Introductory Lectures on Automorphic Forms, Publications of the Math. Soc. of Japan 11, Iwanami Shoten Publishers and Princeton University Press (1971), Chap. 9, Sec. 4.

Borel, A.

[1] Introduction aux Groupes Arithmétiques, Hermann, Paris, 1969.

[2] Linear algebraic groups, Proc. Symp. Pure Math., A.M.S., 9 (1966), p. 10.

[3] Reduction theory for arithmetic groups, Proc. Symp. Pure Math., A.M.S., 9 (1966), pp. 20-25.

[4] Ensembles fondamentaux pour les groupes arithmétiques, Coll. Théorie des groupes algébriques (Bruxelles 1962), Librairie Universitaire, Louvaine; Gauthier-Villars, Paris, p. 37.

Cohen, P. and Sarnak, P.

[1] Discontinuous Groups and Harmonic Analysis, preprint.

Eisenstein, G.

[1] Genaue Untersuchung der unendlichen Doppelproducte, aus welchen die elliptischen Functionen als Quotienten zusammengesetzt sind, und der mit ihnen zusammenhangenden Doppelreihen (Crelle 1847), Mathematische Werke, Chelsea, New York (1975), pp. 357-478.

Fomin, S. V., Gelfand, I. M.

[1] Geodesic flows on manifolds with constant negative curvature (1952), A.M.S. Translations (2), Vol. 1 (1955), pp. 49-65

Gelbart, S., Shahidi, F.

[1] Analytic properties of automorphic L-functions, preprint.

Gelfand, I. M.

[1] Automorphic functions and the theory of representations, Proc. Internat. Congr. of Math., Stockholm (1962), pp. 74-85.

Godement, R.

[1] The decomposition of $L^2(G \setminus \Gamma)$ for $\Gamma = SL(2, \mathbf{Z})$. Proc. Symp. Pure Math., A.M.S., 9 (1966), pp. 211-224.

[2] Introduction à la theorie de Langlands, sém Bourbaki 1966/67, No. 321, pp. 6-9.

[3] The spectral decomposition of cusp forms, Proc. Symp. Pure Math., A.M.S., 9 (1966), pp. 225-232.

Goldfeld, D.

[1] Gauss′ class number problem for imaginary quadratic fields, Bull. Am. Math. Soc. 13 (1985), pp. 23-37.

Harish-Chandra

[1] Automorphic forms on a semisimple Lie group, Proc. Nat. Acad. Sci. U.S.A. 45, pp. 570-573.

[2] Automorphic Forms on Semi-simplie Lie Groups, Lecture Notes in Math., Vol. 62, Springer, New York, 1968.

[3] Harmonic analysis on real reductive groups I, the theory of the constant term, J. of Functional Analysis, Vol. 19 (1975), pp. 102-108.

[4] Spherical functions on a semi-simple Lie group, I, Amer. J. of Math. 80 (1958), pp. 260-261.

[5] The characters of semi-simple Lie groups, Trans. Amer. Math. Soc. 83 (1956), pp. 118-119.

[6] On some applications of the universal enveloping algebra of a semi-simple Lie algebra, Trans. Amer. Math. Soc. 70 (1951), pp. 52-76.

[7] Discrete series for semi-simple Lie groups II, Acta Math. 116 (1966), pp. 18-19.

Hecke, E.

[1] Theorie der Eisensteinschen Reihen höherer Stufe und ihre Anwendung auf Funktionen theorie und Arithmetik (1927), Mathematische Werke, 3 ed., Vandenhoeck & Ruprecht, Göttingen (1983), pp. 461-486.

[2] Analytische Funktionen und algebraische Zahlen, Zweiter Teil (1924), Mathematische Werke, pp. 381-404.

Hejhal, D. A.

[1] The Selberg Trace Formula for PSL(2, **R**), Vol. 2, Lecture Notes in Math., Vol. 1001, Springer, New York (1983), pp. 711-728.

Helgason, S.

[1] Differential Geometry and Symmetric Spaces, Pure and Appl. Math., Vol. 12, Academic Press, New York, 1962.

[2] Groups and Geometric Analysis, Pure and Appl. Math., Vol. 113, Academic Press, New York, 1984.

[3] Fundamental solutions of invariant differential operators on symmetric spaces, Amer. J.Math. 86 (1964), pp. 588-597.

Knapp, A. W.

[1] Representation Theory of Semi-simple Groups, Princeton Math. Series, Vol. 36, Princeton University Press, Princeton, 1986.

Kubota, T.

[1] Elementary Theory of Eisenstein Series, John Wiley and Sons, New York, 1973.

Lang, S.

[1] SL(2, **R**), Addison-Wesley, Reading, Mass., 1975.

Langlands, R. P.

[1] On the Functional Equations Satisfied by Eisenstein Series, Lecture Notes in Math., Vol. 544, Springer, Berlin, 1979.

[2] Review of *The Theory of Eisenstein Systems*, Bull. Amer. Math. Soc. (N.S.) 9 (1983), p. 355

[3] Eisenstein series, Proc. Symp. Pure Math., A.M.S. 9 (1966), pp. 235-252.

Lax, P. and Phillips, R.

[1] Scattering Theory for Automorphic Functions, Annal. of Math. Studies 87, Princeton Univ. Press, Princeton, 1976.

Lehner, J.

[1] Discontinuous groups and automorphic functions, Math. Survey 8. Amer. Math. Soc. (1964), pp. 1-43.

Maass, H.

[1] Über eine neue Art von nichtanalytischen automorphen Funktionen und die Bestimmung Dirichletscher Reihen durch Funktionalgleichung, Math. Ann. 121 (1949), pp. 141-183.

Raghunathan, M. S.

[1] Discrete Subgroups of Lie Groups, Springer, New York, 1972.

Roelcke, W.

[1] Über die Wellengleichung bei Grenzkreisgruppen ersten Art, Sitzber. Akad. Heidelberg, 1956.

[2] Analytische Fortsetzung der Eisensteinreihen zu den parabolischen Spitzen von Grenzkreisgruppen erster Art, Math. Ann. 132 (1956), pp. 121-129.

Rudin, W.

[1] Functional Analysis, McGraw-Hill, New York, 1973.

Schlichtkrull, H.

[1] Hyperfunctions and Harmonic Analysis on Symmetric Spaces, Birhäuser, Boston (1984), Chap. 3.

Selberg, A.

[1] Göttingen Notes, 1954.

[2] Harmonic analysis and discontinuous groups in weakly symmetric spaces with application to Dirichlet series, J. Indian Math. Soc., 20 (1956), pp. 47-87.

[3] Discontinous groups and harmonic analysis, Proc. Internat. Congr. Math., Stockholm (1962), pp. 177-189.

Siegel, C. L.

[1] Advanced Analytic Number Theory, Tata Institute of Fundamental Research, Bombay, 1980.

Terras, A.

[1] Harmonic Analysis on Symmetric Spaces and Applications I, Springer, New York, 1985.

Varadarajan, V. S.

[1] Lie Groups, Lie Algebras, and Their Representations, 2 ed., Springer, New York, 1984.

[2] Harmonic Analysis on Real Reductive Groups, Lecture Notes in Math., Vol. 576, Springer, New York (1977), Part II, Sec. 6.

Warner, G.

[1] Harmonic Analysis on Semi-simple Lie Groups, Vol. I, Sec. 1.2, Springer, New York, 1972.

Weil, A.

[1] Elliptic Functions According to Eisenstein and Kronecker, Springer, New York, 1976.

Yosida, K.

[1] Functional Analysis, Springer, New York, 1971.

Sonderforschungsbereich 170
Mathematisches Institut
D-3400 Göttingen
FRG

MEMOIRS of the American Mathematical Society

SUBMISSION. This journal is designed particularly for long research papers (and groups of cognate papers) in pure and applied mathematics. The papers, in general, are longer than those in the TRANSACTIONS of the American Mathematical Society, with which it shares an editorial committee. Mathematical papers intended for publication in the Memoirs should be addressed to one of the editors:

Ordinary differential equations, partial differential equations and applied mathematics to ROGER D. NUSSBAUM, Department of Mathematics, Rutgers University, New Brunswick, NJ 08903

Harmonic analysis, representation theory and Lie theory to ROBERT J. ZIMMER, Department of Mathematics, University of Chicago, Chicago, IL 60637

Abstract analysis to MASAMICHI TAKESAKI, Department of Mathematics, University of California, Los Angeles, CA 90024

Classical analysis (including complex, real, and harmonic) to EUGENE FABES, Department of Mathematics, University of Minnesota, Minneapolis, MN 55455

Algebra, algebraic geometry and number theory to DAVID J. SALTMAN, Department of Mathematics, University of Texas at Austin, Austin, TX 78713

Geometric topology and general topology to JAMES W. CANNON, Department of Mathematics, Princeton University, Princeton, NJ 08544

Algebraic topology and differential topology to RALPH COHEN, Department of Mathematics, Stanford University, Stanford, CA 94305

Global analysis and differential geometry to JERRY L. KAZDAN, Department of Mathematics, University of Pennsylvania, E1, Philadelphia, PA 19104-6395

Probability and statistics to BURGESS DAVIS, Departments of Mathematics and Statistics, Purdue University, West Lafayette, IN 47907

Combinatorics and number theory to CARL POMERANCE, Department of Mathematics, University of Georgia, Athens, GA 30602

Logic, set theory and general topology to JAMES E. BAUMGARTNER, Department of Mathematics, Dartmouth College, Hanover, NH 03755

Automorphic and modular functions and forms, geometry of numbers, multiplicative theory of numbers, zeta and L-functions of number fields and algebras to AUDREY TERRAS, Department of Mathematics, University of California at San Diego, La Jolla, CA 92093

All other communications to the editors should be addressed to the Managing Editor, RONALD L. GRAHAM, Mathematical Sciences Research Center, AT&T Bell Laboratories, 600 Mountain Avenue, Murray Hill, NJ 07974.

General instructions to authors for

PREPARING REPRODUCTION COPY FOR MEMOIRS

For more detailed instructions send for AMS booklet, "A Guide for Authors of Memoirs." Write to Editorial Offices, American Mathematical Society, P.O. Box 6248, Providence, R.I. 02940.

MEMOIRS are printed by photo-offset from camera copy fully prepared by the author. This means that, except for a reduction in size of 20 to 30%, the finished book will look exactly like the copy submitted. Thus the author will want to use a good quality typewriter with a new, medium-inked black ribbon, and submit clean copy on the appropriate model paper.

Model Paper, provided at no cost by the AMS, is paper marked with blue lines that confine the copy to the appropriate size. Author should specify, when ordering, whether typewriter to be used has **PICA**-size (10 characters to the inch) or **ELITE**-size type (12 characters to the inch).

Line Spacing — For best appearance, and economy, a typewriter equipped with a half-space ratchet — 12 notches to the inch — should be used. (This may be purchased and attached at small cost.) Three notches make the desired spacing, which is equivalent to 1-1/2 ordinary single spaces. Where copy has a great many subscripts and superscripts, however, double spacing should be used.

Special Characters may be filled in carefully freehand, using dense black ink, or **INSTANT** ("rub-on") **LETTERING** may be used. AMS has a sheet of several hundred most-used symbols and letters which may be purchased for $5.

Diagrams may be drawn in black ink either directly on the model sheet, or on a separate sheet and pasted with rubber cement into spaces left for them in the text. Ballpoint pen is not acceptable.

Page Headings (Running Heads) should be centered, in CAPITAL LETTERS (preferably), at the top of the page — just above the blue line and touching it.

LEFT-hand, EVEN-numbered pages should be headed with the AUTHOR'S NAME;

RIGHT-hand, ODD-numbered pages should be headed with the TITLE of the paper (in shortened form if necessary).

Exceptions: PAGE 1 and any other page that carries a display title require NO RUNNING HEADS.

Page Numbers should be at the top of the page, on the same line with the running heads.

LEFT-hand, EVEN numbers — flush with left margin;

RIGHT-hand, ODD numbers — flush with right margin.

Exceptions: PAGE 1 and any other page that carries a display title should have page number, centered below the text, on blue line provided.

FRONT MATTER PAGES should be numbered with Roman numerals (lower case), positioned below text in same manner as described above.

MEMOIRS FORMAT

It is suggested that the material be arranged in pages as indicated below.
Note: Starred items (*) are requirements of publication.

Front Matter (first pages in book, preceding main body of text).

Page i — *Title, *Author's name.

Page iii — Table of contents.

Page iv — *Abstract (at least 1 sentence and at most 300 words).

Key words and phrases, if desired. (A list which covers the content of the paper adequately enough to be useful for an information retrieval system.)

**1980 Mathematics Subject Classification (1985 Revision)*. This classification represents the primary and secondary subjects of the paper, and the scheme can be found in Annual Subject Indexes of MATHEMATICAL REVIEWS beginnning in 1984.

Page 1 — Preface, introduction, or any other matter not belonging in body of text.

Footnotes: *Received by the editor date.
Support information — grants, credits, etc.

First Page Following Introduction - Chapter Title (dropped 1 inch from top line, and centered). Beginning of Text.

Last Page (at bottom) - Author's affiliation.